A Field Guide
to the
Mammals of Egypt

A Field Guide
to the
Mammals of Egypt

Richard Hoath

Illustrations by the author

The American University in Cairo Press
Cairo — New York

To Dr. R.A. Donkin of Jesus College, Cambridge

Copyright © 2003, 2009 by
The American University in Cairo Press
113 Sharia Kasr el Aini, Cairo, Egypt
420 Fifth Avenue, New York, NY 10018
www.aucpress.com

This flexibound edition, with minor revisions, published 2009

Dar el Kutub No. 16765/08
ISBN 978 977 416 254 1

Dar el Kutub Cataloging-in-Publication Data

Hoath, Richard
        A Field Guide to the Mammals of Egypt / Richard Hoath.—Cairo:
    The American University in Cairo Press, 2008
            p.   cm.
            ISBN 977 416 254 4
            1. Mammals—Egypt   I. Title
            599

1 2 3 4 5 6 7 8   14 13 12 11 10 09

Maps by Jeff Miller
Designed by Andrea El-Akshar/AUC Press Design Center
Printed in Egypt

Contents

Acknowledgments

The genesis of this guide came way back in 1989 when I was writing a profile of one of Egypt's leading naturalists, Ibrahim Helmy, for what was then *Cairo Today*. Having patiently endured several interviews for the piece, Ibrahim, with his customary generosity, handed me a copy of the key work on Egyptian mammals he had written with Dale Osborn, both then of the Naval Medical Research Unit No. 3 (NAMRU), *The Contemporary Land Mammals of Egypt (including Sinai)*. Within these pages I discovered the incredible variety of Egypt's mammal fauna and for this gesture, and many others, I owe Ibrahim a huge debt. Also key in opening my eyes to the mammals of not only Egypt but the region has been David Harrison of the Harrison Institute in Sevenoaks in England. Each summer over the past few years, David has opened the doors of his Institute to me and allowed me access to his invaluable collection, without which many of the color plates in this book would not have been possible. He made many useful comments on the plates and has also shared his thoughts on subjects as diverse as the Sinai Leopard and gerbil taxonomy. Thanks also to Paul Bates. Another inspiration has been Sherif Baha El Din, with whom I have had the pleasure of sharing several expeditions, and whose skills as a field naturalist constantly astound. Thanks also to his wife Mindy Baha El Din for her zest and enthusiasm and to Dina Aly and Rafik Khalil for their dedication to Egyptian wildlife. Of the many experts who unselfishly shared their knowledge with me I would particularly like to thank John Grainger of the St. Katherine Protectorate and the wardens who work so hard there under the Egyptian Environmental Affairs Agency (EEAA), and also Tim Wacher. Thanks also to Waheed Salama, also of the

EEAA, and his staff for their help and assistance in the Zaranik Protected Area. My introduction to the sea mammals of the Red Sea was courtesy of Michael Pearson, whose contribution to the conservation of Sinai should never be underrated. Any acknowledgement of the EEAA over recent years must include Salah Hafez and Mustafa Fouda.

I would like to thank the librarians at the British Museum of Natural History and at the Zoological Society of London, especially Ann Sylph, for their patience and assistance.

At the American University in Cairo I owe a huge debt of gratitude to all the faculty, past and present, of the Biology Department for their knowledge and support. The late Ken Horner was my first contact with the department but it is to Derek Russell that I owe an especial debt of thanks for his company and guidance on many desert trips and for sharing his experience and knowledge over many a campfire. Andy Main very kindly checked the manuscript through for taxonomic blips and on several occasions saved me deep embarrassment. I would like to thank him for his thoroughness and stress that any errors that remain are my own. But I would like also to acknowledge here his role in creating one of the most exciting and dynamic departments in the university. I especially thank Jeff Miller, who took on the tedious task of converting my hand drawn range maps into the neat exercises in cartography that now accompany each species entry. Again, any shortfalls are mine. Thanks also to Ian Jenness, Carey Dustin, Moshira Hassan, and Samir Ghabbour.

This guide is unusual in that the writer is also the illustrator, and in the second role I owe a number of special thanks, not least to the mammal curators at the British Museum of Natural History and once more to David Harrison. Special thanks go to Esther Wenman formerly of London Zoo and also to Terry Moore and his colleagues at the Cat Survival Trust, through whom I was able to get unprecedented access to Caracals and Swamp Cats under near natural conditions.

At the American University in Cairo Press I would like to express my gratitude to Arnold Tovell, without whose support this project might have never got off the ground, and to Neil Hewison for his patience over the years when it may have looked as though, having got off the ground, it was never going to land! I am also grateful to Neil for his help with the Arabic names. Thanks also to Matthew Carrington, Sigrún Valsdóttir, my copyeditor Meghan Lynch Schwartz, and designer Andrea El-Akshar.

Many friends and colleagues gave their support over the years spent on the guide. Virginia McKenna of the BornFree Foundation and Julie Wartenberg of WSPA deserve particular mention as do Abdel Aziz Ezz El Arab, Fadel Assabghy, Omar Attum, Dave Blanks, Kate Coffield, Tom Coles, Lillian Craig-Harris, Robin Donkin, Sheryl Ducummon, Mahmoud Farag, the late John Gerhart, Dalia Ghanem, Alan Goulty, Anahid Harrison, Renate Hubinger, Salima Ikram, Chris Magin, Petra Roeglin, Marietta Sawert, Sara El Sayed, Rolf Schmidt, Armin Schröker, Nazli Shafik, and Cluny South. Thanks as ever go to my family.

Preface

Any field guide is the result of an accumulation of knowledge from a wide variety of sources, and this guide is no exception. Innumerable books, articles, and papers have been consulted, the most important of which are listed in the Selected Bibliography. Museum collections have been examined, reports and observations from reliable observers have been compiled, and to this mélange of data has been added my own experience in the field. However, it is important to specifically acknowledge certain key texts, without which this guide would not have been possible.

The first comprehensive work devoted to the contemporary mammals of Egypt was John Anderson's *Zoology of Egypt: Mammalia,* published in 1902. Now very difficult to find, it is a striking tome, and while the text and the taxonomy are inevitably dated, the plates, even a century later, are still stunning for their beauty and their accuracy. However, the landmark text on the mammals of the political entity that is modern Egypt is *The Contemporary Land Mammals of Egypt (Including Sinai)* by Dale Osborn and Ibrahim Helmy, published in 1980. It is this text that is the springboard for the current field guide, and the measurements and ranges of many of the mammal species described in these pages are based on Osborn and Helmy. Measurements have been confirmed whenever possible to museum specimens, and ranges have been adjusted in accordance with more recent research. Even in the relatively short time since 1980, there have been many changes in our understanding of Egypt's mammal fauna. Some of these changes are taxonomic. For instance, while Osborn and Helmy recognized four species of hedgehog from Egypt, most authors now recognize just two. The classification of the shrews, gerbils, and gazelles is particularly open to debate, and the current author has generally followed Nowak (1999 ed.) in these instances. At the more contentious level of subspecies, again the work of Osborn and Helmy has generally been followed. Other changes relate to recent research. Since Osborn and Helmy, certain additional species have been added to Egypt's mammal fauna, for instance the Marbled Polecat. An extremely valuable text for species in Sinai, or species very rare in Egypt but more widespread in the Middle East has been *The Mammals of Arabia* (1991 ed.) by David Harrison and Paul Bates. For anyone needing information on skull and dental characteristics and other information beyond the scope of a field guide, these texts are invaluable and both include extensive bibliographies that take the researcher back to primary sources. Full bibliographic data on these texts can be found in the Selected Bibliography.

Osborn and Helmy did not include the bat fauna of Egypt, and in this case two key texts have been used as a starting point for this guide. The first is the aforementioned Harrison and Bates, the second is Qumsiyeh's *The Bats of Egypt* (1985). It is from these texts that the key measurements for Egyptian bat species have been taken, confirmed whenever possible to museum specimens.

Marine mammals, the cetaceans and the sirenians, were not covered by Osborn and Helmy, and in this instance records have been found by going back into the scientific literature or by the sight records (in at least one instance supported by photographs) of reliable observers. In the Red Sea it has been possible to use *Key Environments: Red Sea* (1987) by Edwards and Head, published in collaboration with the IUCN. The authors of this text note in their introduction to the cetaceans that, "despite their relatively large size, our knowledge of cetaceans in the Red Sea is extremely fragmentary." With the dramatic increase of tourism in the Egyptian Red Sea, and in particular dive tourism often on live-aboard boats, the number of records of whale and dolphin species should increase. For this reason, a number of species confirmed by Edwards and Head from the southern Red Sea and possible but not confirmed from Egyptian waters, have been included in the plates.

Seminal though the scientific texts are for anyone claiming expertise in the mammal fauna of Egypt, the necessarily academic prose and the proliferation of scientific minutiae is frequently offputting to a wider audience, an audience it is becoming increasingly important to access. Egypt's mammal fauna is everywhere threatened by an ever-expanding human population, by direct hunting and by rampant urbanization. Nowhere is this clearer than in the proliferation of tourist developments along the Mediterranean and Red Sea coasts.

The wholesale destruction of the Mediterranean coast and the oft-quoted aim of creating a "Red Sea Riviera" might seem to cast an ominous shadow over the future of these environments. The explosion of four-wheel drive tourism from these new tourist cities indeed represents a serious threat to a fragile desert environment. Yet potential catastrophe is also potential opportunity. A desert where all there is to see is desert and the myriad of tracks of previous 'pioneers,' will attract no one. A vibrant, living desert, a desert with jirds, gerbils, and jerboas, with foxes and hares, hyraxes, ibexes and gazelles, will be an environmental asset to preserve and treasure. We must work to prevent the former scenario and to encourage the latter. Anyone fortunate enough to have experienced a first encounter with wild gazelles in the vastness of the Egyptian desert will know exactly what I mean.

This is where this Field Guide comes in. While it must retain the scientific rigor and accuracy to be a valuable tool for the expert, it is also a celebration of the diversity of Egypt's mammal fauna aimed at a new and growing audience of visitors, foreign and local, to Egypt's wilder places at sea and on land. Gone are the black-and-white photographs of carefully preserved museum cadavers, and for the first time all of Egypt's mammals are accurately portrayed, mainly in color, as alive and alert as in nature. The guide is designed to fill a growing need among this new generation of visitors and explorers and to foster and encourage further recruits to Egypt's growing environmentalist movement.

Egypt

The Biogeography of Egypt

The political entity of Egypt makes up the northeasternmost corner of the African continent, together with the Sinai Peninsula. It has an area of c. 1,019,449km^2, of which around 18,000km^2 are administered by Sudan (Sudan Government Administration Area). The land borders to the north and east are clearly defined by the coastal areas bordering the Mediterranean and Red Seas respectively, but to the west and south, as well as the eastern land frontier, the borders follow no natural boundaries but are colonial legacies; to the south with Sudan, the west, Libya, and the east, Palestine and Israel. While beyond the immediate concerns of zoologists, these artificial boundaries are important since they explain the main reason why Egypt has relatively few endemic plant and animal species. Endemic plant species are highlighted in Boulos (1999 and 2000). Amongst the animals there are, for example, two endemic species of butterfly, *Pseudophilotes sinaicus* and *Satyrium jebelia*, the endemic Kassas's Toad *Bufo kassasii*, and the mammals, Flower's Shrew *Crocidura floweri*, the Pallid Gerbil *Gerbillus perpallidus*, Egyptian Weasel *Mustela subpalmata,* and possibly Mackilligin's Gerbil *G. mackilligini*. There are no endemic bird species, again largely a function of Egypt's unnatural frontiers.

Within these borders, Egypt is predominantly arid, the often-quoted figures being between 95 and 96% desert. The remaining percentage is largely made up of the Nile Valley and Delta, the latter expanding out fanlike from just north of Cairo to the Mediterranean. The Nile originates far south of Egypt's southern frontier with Sudan; the White Nile rising in the highlands of central Africa and the Blue Nile in the mountains of Ethiopia. It is the latter that fuels the river with the bulk of its water, and in the past (before the waters were tamed by the various dams and barrages), it was the floods of the Blue Nile waters that dictated the inundations of the Nile in Egypt. These inundations provided the agricultural lands of the Valley and Delta with an annual supply of fertile silt that was crucial to its fertility throughout over 5,000 years of human agricultural activity. It is probably true to say that no other area on earth has been subject to such a long and intensive invasion of human activity. As will be seen, it is now a wholly artificial environment. There are no natural tributaries of the Nile within Egypt's borders today, though the Western Desert oases of Kharga, Dakhla, Farafra, and Bahariya trace a long, fossil, subterranean course of the Nile and provide outposts of fertility in an otherwise extremely dry region of the planet.

The importance of climate can be seen from the accompanying maps. Both rainfall and temperature alter radically in a very short geographical distance as one moves away from the north coast. The narrow strip of cooler, wetter desert along the north coast supports a similarly narrow biome, including plants and animals that owe more to Mediterranean conditions than Saharan. Even within this zone, climatic and geological features support plant and animal species not found elsewhere in Egypt, notably in the extreme west and east. To the south, precipitation, often in the form of snow, increases in the mountains of South Sinai, while there is an increase in terms of orographic rainfall in the Gebel Elba region in the southeasternmost corner of the country. However, through much of Egypt the pattern is broadly similar, temperature rising and rainfall declining rapidly inland from the northern coast and then more steadily south over the rest of the country.

Within this general pattern, the Egyptian landscape varies dramatically and with it, the plants and animals it supports. While microclimatic conditions will always vary on the smallest scale, a number of distinct biological regions can be noted in Egypt. Sweeping attempts to catalog its flora and fauna within four biogeographical regions; namely, the Saharo-Sindian, Irano-Turanian, Mediterranean, and Afro-tropical are in danger of oversimplifying the impact of historical and relatively recent geo-climatic influences and, most recently of all, the impact of humans—an impact that simply cannot be ignored, especially over the Nile Valley and Delta. For example, the Nubian Ibex *Capra ibex* probably evolved from Eurasian relatives of the African antelope group. In Eurasia, this stock evolved in competition with the deer (Cervidae) to become adapted to marginal montane habitats. During the last ice ages, these caprines reinvaded northern Africa, but with the retreat of the colder climate were left in the mountains of northeastern Africa, extending down to an isolated area in the Ethiopian Highlands. Today, this seemingly ancient, yet relatively recent, distribution leaves the Nubian Ibex as a resident of the Sinai mountains and the highlands of the Red Sea Mountains. Within historical times, and especially the last two hundred years, the influence of humans is sadly evident in the distribution of the Nubian Ibex. Due to hunting and habitat disturbance, it has become confined within its already limited habitat to those areas most remote and most inaccessible to human disturbance. Little of Egypt, and its flora and fauna, has been left untouched by human hand.

Historically, all evidence indicates that between 8000 and 3000 BC, this northeastern corner of the African continent underwent a climate change, becoming both hotter and drier. Petroglyphs made by the ancients indicate a fauna much more akin to the East African savanna fauna of today. For instance, at Silwa Bahari there are predynastic rock drawings of African Elephants *Loxodonta africana*, White Rhinos *Ceratotherium simum*, and Gerenuk *Litocranius walleri*, as well as hunting scenes with Ostriches *Struthio camelus* being pursued with bows and arrows. In the rock tombs of the pharaohs, there are frequent representations of the Bubal Hartebeeste *Alcelaphus buselaphus buselaphus*, a subspecies now extinct but related to the hartebeestes of current East and southern Africa. Along the Nile, the Hippopotamus *Hippopotamus amphibius* survived until historical times.

However, to define Egypt today loosely as a hot, desert country is once more to oversimplify: the different types of desert differ radically in topography and resultant flora and fauna. The Nile Valley and Delta apart, the Egyptian desert regions are represented by distinct plant and animal communities. What follows is an overview of these regions with a summary of the Protected Areas designated within each biome.

The Northern Coastal Strip

Stretching from the border with Libya to Alexandria, the coastal desert's distinctive feature is the relatively high, and more consistent, rainfall and low temperature compared to the rest of Egypt. As can be seen from the map, the rainfall decreases very rapidly inland from the coast, giving this zone a maximum width of around 50km along its 600km length. The distinctive geography of this narrow coastal strip allows it to play host to Egypt's most prolific flora, both in terms of absolute number and of species diversity. Unsurprisingly, this rich flora supports a wide range of animal life. Distinctive birds include the Barbary Partridge *Alectoris barbara* (probably locally extinct), Houbara Bustard *Chlamydotis undulata*, Dupont's Lark *Chersophilus duponti*, Thekla Lark *Galerida theklae*, Temminck's Horned Lark *Eremophila bilopha,* and Red-rumped Wheatear *Oenanthe moesta.* Characteristic mammals include the Long-eared Hedgehog

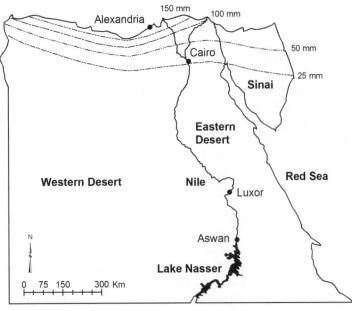

Annual rainfall in Egypt in mm (after Osborn and Helmy [1980])

Hemiechinus auritus, Cape Hare *Lepus capensis,* Anderson's Gerbil *Gerbillus andersoni,* Shaw's Jird *Meriones shawi,* Fat Sand Rat *Psammomys obesus,* Lesser Molerat *Spalax leucodon,* Middle Eastern Dormouse *Eliomys melanurus,* Greater Egyptian Jerboa *Jaculus orientalis,* and Four-toed Jerboa *Allactaga tetradactyla.* Sadly, this coastal strip is also one of the most threatened habitats. Tourist developments expanding west from Alexandria have destroyed much of this habitat to Mersa Matruh and threatened expansion west will probasbly mean that no area east of Sallum is safe. With the exception of the al-Omayed Biosphere Reserve inland from al-Alamein where the Red-rumped Wheatear may still breed, no area within this coastal desert is protected.

One Protected Area: al-Omayed Biosphere Reserve.

The Western Desert

The vast expanse of Egypt west of the Nile Valley and south of the north coast is collectively known as the Western Desert, the northeasternmost portion of the Sahara Desert. The area is characterized by low relief and areas of vast, inhospitably arid *hamada* and sand plains with very little rainfall. This barren landscape, in parts virtually lifeless, is relieved by areas of massif, such as Gebel Uweinat in the southwest, and the oases scattered along fossil watercourses. Foremost among these are Siwa, Bahariya, Farafra, Dakhla, Kharga, and Wadi Natrun, which show a fauna similar to that of the Nile Valley and Delta. The Fayoum is not strictly speaking an oasis but rather nowadays a dead-end branch of the Nile. It too has a fauna similar to the Nile Valley and Delta.

To the north of the Western Desert, the dominant physical feature is the Qattara Depression covering 19,500km^2 and reaching below sea level 134m in depth. The floor of the depression too has isolated oases, sufficient to support *Acacia* species and a number of salt lakes. The western region of the Western Desert is marked by the dunes of the Great Libyan Sand Sea, while to the southwest the dominant features are the Gilf al-Kebir and the isolated massif of Gebel Uweinat. Typical birds of the region include Spotted Sandgrouse *Pterocles senegallus,* Hoopoe Lark *Alaemon alaudipes,* and White-crowned Black Wheatear *Oenanthe leucopyga.* The mammal fauna of the Western Desert is

sadly depleted. Within the last two hundred years or so, the Bubal
Hartebeeste, Addax *Addax nasomaculatus*, and Scimitar-horned Oryx
Oryx dammah have all disappeared. The Slender-horned Gazelle
Gazella leptoceros is much depleted and it is doubtful whether the
Cheetah *Acinonyx jubatus* still clings on. Typical mammal species extant
in the region include Lesser Egyptian Jerboa *Jaculus jaculus*, Giza Gerbil
Gerbillus amoenus, Rüppell's Sand Fox *Vulpes rueppelli*, and Dorcas
Gazelle *Gazella dorcas*. The Gilf al-Kebir and Gebel Uweinat repre-
sent the last strongholds in Egypt of the Barbary Sheep *Ammotragus
lervia*. The Western Desert's only Protected Area covers just 1km^2.

Four Protected Areas: Hasana Dome Protected Area, Gilf Kebir
National Reserve, White Desert Protected Area, and Siwa
Protected Area.

The Eastern Desert

The Eastern Desert is very different from its western counterpart. It
broadly consists of a range of sedimentary mountains that separate
the Nile Valley from the Red Sea, the northernmost extension of
which are the Muqattam Hills east of Cairo. A coastal plain of variable
width separates these mountains from the Red Sea. These ranges are
dissected by a series of deep wadis that reflect a time when water was
the dominant erosion agent in these uplands. Two examples include
Wadi Hof, south of Maadi, and Wadi Rishrash, north of Beni Suef.
These wadi floors are still often vegetated, in contrast to the barren
plateaus, there being sufficient groundwater to support such species
as *Acacia* spp. and Tamarisk *Tamarix nilotica*. The wadi walls are often
precipitous, and dry waterfalls are a common topographical feature.
While the Eastern Desert is classified as hyper-arid, when rain does
fall it can be torrential and lead to flash floods that are still the dominant erosional force in the region. The plateau tops are extremely arid.

The fauna of the Eastern Desert is strikingly different from that
of the Western Desert, reflecting the very different topography and
the importance of the Nile as a zoogeographical barrier. Birds such
as the Sand Partridge *Ammoperdix heyi*, Scrub Warbler *Socotocerca
inquieta*, White-crowned Black Wheatear, and Mourning Wheatear
Oenanthe lugens are typical of these deserts. The mammal species
show a close parallel to those of South Sinai including the Golden

Spiny Mouse *Acomys russatus*, Bushy-tailed Jird *Sekeetamys calurus*, Nubian Ibex, and Striped Hyena *Hyaena hyaena*. Recent records appear to confirm this connection with South Sinai with records of Hume's Tawny Owl *Strix butleri* and Blanford's Fox *Vulpes cana* from the Eastern Desert. An exception is Rüppell's Sand Fox, which is found in both deserts. To the west, clearly the Nile is not a complete barrier, based on the distribution of the Lesser Egyptian Jerboa and the Lesser Egyptian Gerbil *Gerbillus gerbillus*.

Six Protected Areas: Petrified Forest Protected Area, Wadi Degla Protected Area, Sannur Cave National Monument, Wadi al-Asyuti Protected Area, Wadi Allaqi Protected Area, and Wadi Gimal Protected Area.

The Sinai Peninsula

The Sinai Peninsula is of immense importance in any discussion of the flora and fauna of Egypt. It is an inverted triangle of land, some $61,000 \text{km}^2$ in area, with a northern shoreline on the Mediterranean and its southern sides defined by the Gulfs of Aqaba and Suez of the Red Sea. It is the land connection between Africa and Asia and, at the same time, the land barrier. It is the land connection because in pre-historic times, i.e., until ten thousand years ago, the climate was such that there was a continuous band of non-desert vegetation across Sinai connecting Asia Minor with the Nile Valley. The zoogeographical traces of this connection can be seen in the Swamp Cat *Felis chaus* and the Bandicoot Rat *Nesokia indica*. It is the land barrier because, after this time period, the arid wastes of Sinai represented a barrier to any species colonizing the region. Thus, the Swamp Cat and the Bandicoot Rat have become isolates, species of Asiatic origin now separated from their congeners by the relatively recently barren wastes of Sinai.

Sinai, however, cannot be considered as a whole since the north and south of the peninsula are very different. The south can be best represented as a continuation of the Eastern Desert, both geological-ly and zoogeographically. It too is characterized by mountainous ter-rain dissected by water-eroded wadis. Many faunal species are com-mon to both zones. Amongst the birds these include the two wheatears, the Mourning and the Hooded *Oenanthe monacha*, Sand Partridge, Scrub Warbler, and, perhaps (it may not still survive in

Sinai), Lammergeier *Gypaetus barbatus*. Mammals too bridge the two regions. Those typical of both include the Golden Spiny Mouse, Bushy-tailed Jird and, most recently (in terms of discovery), Blanford's Fox. But no mammal better emphasizes the similar nature of the terrain than the Nubian Ibex.

However, Sinai is also characterized by species found nowhere else in Egypt—a result of the isolationist factors mentioned above. Two butterflies, *Pseudophilotes sinaicus* and *Satyrium jebelia,* are endemic to the region. A number of reptile species, including the Sinai Banded Snake *Coluber sinai*, Crowned Peace-Snake *Eirenis coronella,* and a subspecies of Ornate Dabb Lizard *Uromastyx ocellatus ornatus*, are restricted in Egypt to the Sinai Peninsula. Amongst the birds, the list is longer and includes the Sinai Rosefinch *Carpodacus synoicus*, Palestine Sunbird *Nectarinia osea*, Arabian Babbler *Turdoides squamiceps*, and Yellow-vented Bulbul *Pycnonotus xanthopygos*. Amongst the mammals, there is one possible (on taxonomic grounds) endemic subspecies, namely, the South Sinai Hedgehog *Paraechinus aethiopicus dorsalis*. Similarly debatable on taxonomic grounds is the Sinai Leopard *Panthera pardus jarvisi,* now probably extinct. There are no mammal species endemic to Sinai.

North Sinai is characterized by low rolling sand dunes, very different from the mountains of the south. From North Sinai, the Negev Jird *Meriones sacramenti* and Tristram's Jird *Meriones tristrami* are found nowhere else in Egypt, while the Fennec Fox *Vulpes zerda* and the Sand Cat *Felis margarita* have also been recorded. Of most interest, however, is the very northeastern corner of Sinai where, as mentioned above, rainfall is higher and a number of faunal species creep across the border from Palestine and Israel. Foremost amongst the amphibians must be the Tree Frog *Hyla savignyi*, which can be looked for at the base of date palms near Rafah. Birds such as the Syrian Woodpecker *Dendrocopos syriacus* and Great Tit *Parus major* now breed, and mammals such as the recently confirmed Marbled Polecat *Vormela peregusna* and a porcupine species *Hystrix* sp. are now known to occur.

One National Park: Ras Muhammad National Park.

Six Protected Areas: Zaranik Protected Area, al-Ahrash Reserve, St. Katherine Protected Area, Nabq Protected Area, Abu Galum Protected Area, Taba Protected Area.

Gebel Elba

The rocky massif of Gebel Elba in the very southeasternmost part of Egypt is of great faunal significance. Rainfall, largely orographic, makes this region far less hyper-arid than the Eastern Desert to its north. The richer vegetation dominated by *Acacia* spp. and *Euphorbia* spp. supports a fauna more akin to the Afro-tropical region than to the Palearctic. A significant proportion of Egypt's butterfly species are found nowhere else but in Gebel Elba such as the stunning *Colotis danae* and *Charaxes hansali*. Much of the area has yet to be properly studied, but the bird fauna alone serves to indicate the African nature of the biome. Rosy-patched Shrike *Rhodophoneus cruentus*, Shining Sunbird *Nectarinia habessinica*, and Fulvous Babbler *Turdoides fulvus* are just three examples. Amongst the mammal species that are recorded only in Gebel Elba are the Aardwolf *Proteles cristatus*, Zorilla *Ictonyx striatus*, and Small-spotted Genet *Genetta genetta*, all of the Afro-tropical faunal community. Wild Ass *Equus africanus* may still exist in the area.

One Protected Area: Gebel Elba Protected Area.

The Nile Delta and Valley

It is all too common, particularly in the flowery language of the tour guide, to describe the Nile Delta and Valley as 'timeless' and 'unchanging,' yet this is far from the truth. Few parts of the world have been subjected to such prolonged and intensive human influence, and the present day Nilotic environs are a product of this influence. A cursory glance at the friezes in many of the tombs of the ancients serves to support this stance. Often the pharaoh or noble is portrayed hunting in a swamp of Papyrus *Cyperus papyrus* for Hippopotamus and Nile Crocodile *Crocodylus niloticus*. The former is now extinct in Egypt and the latter, since the late nineteenth century, is only found south of the Aswan High Dam. More recently, the Wild Boar *Sus scrofa* disappeared in 1912. These papyrus swamps are now entirely gone and the only remnant of the original Nilotic vegetation in Egypt now exists on the islands between Aswan and the Old Dam. This is now a Protected Area.

Today, the Delta and Valley of the Nile support an almost entirely exotic flora courtesy of modern agriculture. One need not look any further than the crops grown today. Major crops such as cotton, tomatoes, potatoes, sweet corn (maize), and sugar cane all originate in the Americas. Virtually all the trees seen in the urban environments are exotic and, in the rural areas, the ubiquitous *Eucalyptus* spp. are an import from Australia. In short, the ancient Egyptian would find the current Nile flora virtually unrecognizable. Familiar species such as Papyrus and Lotus *Nymphaea lotus* are virtually extinct in the wild.

The current fauna of the Nile Delta and Valley includes the endemic Kassas's Toad and the Egyptian Square-marked Toad *Bufo regularis*. Amongst the birds there are a number of species of African origin (for which the Nile has acted as a corridor north) such as the Senegal Coucal *Centropus senegalensis*, Senegal Thick-knee *Burhinus senegalensis*, and Black-shouldered Kite *Elanus caeruleus*. Other typical species include the Common Bulbul *Pycnonotus barbatus*, Graceful Warbler (Prinia) *Prinia gracilis*, and Painted Snipe *Rostratula benghalensis*. Typical mammals include the Egyptian Mongoose *Herpestes ichneumon*, Striped Weasel *Poecilictus libyca*, Nile Kusu *Arvicanthis niloticus*, and the endemic Flower's Shrew. Some mammals typical of the Delta appear to be African outposts of predominantly western Asiatic distributions, reflecting a cooler, wetter period where the Sinai was not a desert barrier. These include the Swamp Cat and the Bandicoot Rat. Mention should also be made of the commensals characteristic of the agricultural and urban areas. These include the House Mouse *Mus musculus*, Brown Rat *Rattus norvegicus*, and House Rat *Rattus rattus*. It is interesting to note that the Weasel *Mustela nivalis* is an almost entirely urban animal in Egypt.

Above the Aswan High Dam, the Nile Valley has been inundated to form Lake Nasser. The shores of this lake are, for the most part, barren. However, the lake does support Egypt's only population of Nile Crocodiles and probably the last remaining Nile Soft-shelled Turtles *Trionyx triunguis*. Egyptian Geese *Alopochen aegyptiacus* are common and Afro-tropical species such as Yellow-billed Stork *Mycteria ibis*, African Skimmer *Rynchops flavirostris*, and Pink-backed Pelican *Pelecanus rufescens* are regularly recorded. Jackals *Canis aureus* and Red Foxes *Vulpes vulpes* are regularly recorded from the shores and from the islands that emerge and disappear as the lake levels sink and rise.

Six Protected Areas: Saluga and Ghazal Protected Area, Ashtum al-Gamil Protected Area, Wadi al-Rayyan Protected Area, Lake Qarun Protected Area, Lake Burullus Protected Area, and Nile Islands Protected Area.

Marine Environments

Egypt has two coastlines, one bordering the Red Sea, extending north to the Gulfs of Aqaba and Suez, and one bordering the Mediterranean. Formerly discrete, these two marine regions are now connected by the Suez Canal though it is still unclear how much interchange there is between the biomes. Certainly there is no evidence that sea mammals are using the channel.

The Red Sea is continuous with the Indian Ocean through the narrow Bab al-Mandab between Yemen and Djibouti. Further north, the Gulfs of Aqaba and Suez are very different in character: the former being much deeper than the latter and supporting richer coral reefs. The marine fauna of the Red Sea is essentially Indo-pacific and includes such warm water species as the Pantropical Spotted Dolphin *Stenella attenuata*, Spinner Dolphin *Stenella longirostris*, and Short-finned Pilot Whale *Globicephala macrorhynchus*. There are very few records of the great whales from the Red Sea and no records of any of the beaked whales. It may be that the shallow water that marks the Bab al-Mandab is a barrier to these deep-water species. The Dugong *Dugong dugon* is still found, though in much reduced numbers.

The Mediterranean is also virtually a closed sea, connected to the Atlantic by the narrow Straits of Gibraltar. Its waters are too cold for reef-building coral species. As with the Red Sea, there are very few records of the great whales in the Egyptian Mediterranean and no confirmed records of the beaked whales, though Cuvier's Beaked Whale *Ziphius cavirostris* could conceivably occur. The Mediterranean Monk Seal *Monachus monachus* was found along the Egyptian north coast but there have been no records of this species since 1921. With the wholesale development of the Egyptian Mediterranean coastline, it is highly unlikely that this species will return, though there is reportedly a very small population in Libya.

Protected Areas: Each of the coastal southern Sinai Protected Areas extends into the marine environment. The Mediterranean marine environment has no Protected Areas though the lagoon at Zaranik is saline and Lake Bardawil is Egypt's only designated RAMSAR site. Northern Red Sea Islands Protectorate.

Current Threats and the Status of Egyptian Mammals

The present Egyptian mammal fauna is sadly a much-depleted one. As has been mentioned, within the last two hundred years several species, such as the Addax and Scimitar-horned Oryx, have disappeared completely while many others, especially the larger mammals, have suffered declines. Among the smaller mammals, the Four-toed Jerboa is a cause for concern and several of the bat and shrew species are known from very few specimens. Others such as the Zorilla and the Aardwolf are at the very edge of their range in Egypt and are probably naturally rare.

The most obvious threat to the larger mammals, such as the Nubian Ibex and the gazelle *Gazella* species, is that of direct hunting. All these species are protected under Egyptian law, but the law is rarely enforced. This hunting is carried out by locals as well as foreign sport hunters, particularly from the Gulf. While hunting in certain strongholds of these species (such as the mountains of South Sinai) has now been controlled with the declaration of an extensive network of Protected Areas, in other, more remote regions it still takes place, such as in the southern Eastern Desert. With the opening up of the coastal regions of the Eastern Desert for tourism, these areas will become more and more accessible causing concern for the local wildlife.

In a country with a rapidly growing population and finite resources, there is inevitable pressure on the environment and habitat destruction is a major threat to many mammal species. Nowhere is this more apparent than along the north coast and the Mediterranean coastal strip. This fragile habitat is disappearing rapidly as tourism advances inexorably and as concrete tourist village after concrete tourist village is constructed. Today, the entire coast from Alexandria to al-Alamein is developed and the development is creeping west at an alarming rate. Quarrying and mining also threaten this habitat. None of this coastline is protected. Inland, ill-planned agricultural development has turned what used to be relatively fertile semi-desert into true desert, as the shallow subsoil is loosened and exposed by plowing. The result is real desert good for neither agriculture nor wildlife. The coast of North Sinai suffers in the same way. The Zaranik Protected Area west of al-Arish at least preserves some of the natural habitat, though even here the vegetation is being denuded by overgrazing. Along the Red Sea coast, tourist development is proceeding at an alarming rate with new centers being developed south of Hurghada at Safaga, Quseir, and Marsa Alam. Except for the recently declared Wadi Gimal Protectorate, there are no Protected Areas along this coast until the Elba Protectorate in the very southeastern part of the country. In all these tourist areas, the opening up of the deserts allows access to four-wheel drive vehicles that can be harmful in churning up the sand, destroying the vegetation, and exposing—thus killing off—dormant seeds, thus reducing future pasture. This has been of particular concern in parts of South Sinai.

Inland, the expansion of the cities into the desert areas and the building of new cities is threatening desert habitat. It was because of these threats that the Wadi Digla Protected Area was declared in 1999. The results of these urban encroachments can be seen along all the main roads leading out of Cairo. An interesting aspect of these urban encroachments is its apparent effect on the distribution of the Red Fox. This species seems to be expanding along the roads following the development into new areas where it seems to out-compete the desert Rüppell's Sand Fox. In Osborn and Helmy (1980) the Red Fox was virtually unrecorded from Sinai. Today, it can be found as far south as the Ras Muhammad National Park.

In the Delta and Nile Valley, pollution from waste disposal and the use of toxic pesticides and herbicides has probably had an effect on the mammal fauna. This may not necessarily be negative in pure mammal terms (but certainly in ecological terms!). The documented decline in birds of prey such as the Black Kite *Milvus migrans* and the Black-shouldered Kite may well have resulted in an increase in the numbers of their prey species, such as the Cairo Spiny Mouse *Acomys cahirinus* and the Nile Kusu. In marine environments, especially the Red Sea extending up the Gulf of Suez, oil pollution may have an effect on marine mammals.

Just how rare many of Egypt's mammals are is in many cases unclear since very little survey work has been done. Hence, although species like the Lesser White-toothed Shrew *Crocidura suaveolens* and the Pygmy Pipistrelle *Pipistrellus ariel* are known from very few Egyptian records, they may prove to be underrecorded rather than actually rare in Egypt. Others, such as the Sinai Leopard, are clearly very rare if indeed extant. Some idea of the status of certain mammal species can be discerned by whether it is listed in the Convention on the International Trade in Endangered Species (CITES), to which Egypt is a signatory, or listed by the World Conservation Union (IUCN). It should be noted that an international convention signed by Egypt becomes Egyptian law. CITES species are listed on one of three appendices, the first two of which are relevant to Egypt. Those on Appendix I include "all species threatened with extinction which are or may be threatened by trade." Trade in these species is only "authorized in exceptional circumstances." Appendix II lists all those species not currently threatened with extinction by trade but "may become so unless trade in specimens of such species is subject to strict regulation." All Egyptian species listed on these two appendices are documented as so under status in the species description.

The IUCN listing catalogs the degree to which an animal, or plant, is endangered according to certain criteria. The following categories are used.

Extinct: Extinct or close to extinction.

Endangered: Near to extinction and likely to become extinct unless action is taken.

Vulnerable: Unless action is taken likely to move into the endangered category in the near future.

Indeterminate: Known to be in one of the three categories above but there is too little information to assign to a specific category.

Insufficiently Known: Thought to be in one of the above categories but there is too little known to assign to a specific category.

All taxa listed by the IUCN are published in their Red Data Books. Where an Egyptian mammal is listed under any of the above categories, this is documented under status in the species description. For a full discussion of the international and national legislation relevant to Egyptian wildlife, see Baha El Din (1999).

The Future

With so many pressures from every side, it would be easy to dismiss the chances of much of Egypt's wildlife—and the mammals in particular—surviving very long. Tourism spreads with all its consequences, population grows, urban areas expand, and pollution multiplies. The scenario indeed looks bleak. Where there are laws or international treaties and agreements, enforcement is often lax. Public education is minimal and little is being done to change the preconceptions of people regarding wildlife. No one need look further than the desperate conditions and minimal educational attempts at the Giza Zoo to see that.

However, there are positive developments. In the past twenty years, at least twenty-seven National Parks/Protectorates have been declared. While many need far more financing and resources, others are performing their function of protecting the natural environment. While by no means trouble free, the Protectorates of South Sinai show what can be done—albeit with foreign funding, in this case by the EU. These areas have not only received extensive funding, but also extensive publicity. Ras Muhammad is not only a National Park, it is also world renowned as a dive site and an area of biodiversity of global importance. Sadly, many other such areas receive far less recognition and are open to abuse and violation.

It is of vital importance for a country that treasures its historical heritage so highly that its natural heritage is valued and protected with equal vigor. Whereas the pyramids of Giza, the tombs and temples of Thebes, and other historical sites up and down the Nile Valley receive great attention as part of Egypt's historical legacy, the same cannot be said for its far, far older and more natural flora and fauna. Too much has been lost already as the coming pages will testify. It is sad to relate that many of the animals and birds held sacred by the ancients are now extinct in modern Egypt: the Sacred Baboon *Papio hamadryas*, the

Sacred Ibis *Threskiornis aethiopicus*, the Lion *Panthera leo*, the Crocodile (north of the High Dam), and the Hippopotamus. Yes, climate change has had a role, but one cannot simply ignore human agency.

What is lacking at the moment is a young generation coming forward to treasure what is left and to preserve it. Too often in the recent past there has been talk of captive breeding programs and re-release back into the wild. This is both very expensive and impractical. Captive breeding programs (for species such as the Barbary Sheep) can only work if there are sufficient areas of natural habitat for bred animals to be released back into, if sufficient resources are available, and if, on being released, the safety of the released animals can be assured. None of these guarantees currently exist. The future of what is left of Egypt's mammals will be best assured by protecting what is left of those species in the wild and assuring the safety of what remains of their habitats. Anything else will be an irresponsible and expensive gamble.

Using the Guide

This guide is designed to help the reader identify mammals seen in the field in Egypt. Sometimes this can be straightforward. A rabbit-like mammal with very long ears and a short, black-and-white patterned tail is clearly a Cape Hare *Lepus capensis,* while a goat-like mammal with knobby, back-curved horns and strikingly patterned limbs seen in the mountains of South Sinai can safely be identified as a Nubian Ibex *Capra ibex,* once the domestic goat is eliminated. However, very often things are not so clear-cut. Mammals are often very difficult to observe. They are often very shy and secretive, and while less than 3% of all bird species are nocturnal, making things relatively straightforward for the ornithologist (in Egypt, the percentage is 2.61%, using Goodman and Meininger [1989]), many mammal species, including most of the difficult-to-identify rodents and bats, are only active at night. And as most of the prey species are active by night, so are the predators.

The first step is to try and match the mammal seen to the plates, taking note of distinguishing features, prominent markings, and relative size. Each individual plate has all the species represented to scale, and actual size can be found by referring to the text. Proportions, particularly those of the tail, hind feet, and ears can all be of importance. The summarized text opposite each plate emphasizes the key identification points. Thus, a desert rodent bounding across the road at night on its hind legs, with a long, black and white tufted tail can be narrowed down to one of the jerboa species from the plate (*Allactaga* sp. or *Jerboa* spp.).

The second step is to check the ranges of the likely species to the maps adjacent to the individual species' texts. These maps show the range within which each species may be found, assuming suitable habitat. Ranges expand and contract due to natural and human factors, so range alone cannot firmly identify a mammal but can at least highlight the most likely species. Thus, the jerboa cited above is almost certain to be the Lesser Egyptian Jerboa *Jaculus jaculus* in the Eastern Desert. In Sinai, it must be compared with the Greater Egyptian Jerboa *Jaculus orientalis* while along the northern coastal strip the Four-toed Jerboa *Allactaga tetradactyla* is also a possibility. That said, many species, especially some of the bats and sea mammals, may enjoy wider ranges than have been indicated, as lack of research may explain the relatively few records of certain species rather than natural scarcity.

Third, refer to the text. Any other information such as habits and habitat, associated species, or even time of day observed can be important. For example, a mouse seen in rocky habitat in the Eastern Desert by day is more likely to be the Golden Spiny Mouse *Acomys russatus* than the Cairo Spiny Mouse *Acomys cahirinus*, which is largely nocturnal. Similarly, a robust gerbil-like rodent seen in a salt marsh habitat by day is more likely to be a Fat Sand Rat *Psammomys obesus* than one of the jirds *Meriones* spp. Secondary details may also be useful such as any tracks or spoor and, for this purpose, some of the more distinctive tracks have been illustrated after the species plates. Cranial characteristics and detailed dental formulae are beyond the scope of a field guide but can be found in the literature detailed in the Selected Bibliography.

The Text
Each mammal species has its own entry giving the following details.

Species name: The most widely used English name is given in bold with alternative names following in brackets. The scientific name is then given for the most part following Nowak (1999), but see below for taxonomic note. Subspecies, where relevant, generally follow Osborn and Helmy (1980). Finally, the Arabic name or names are given where possible, following Wassif (1995). For ease of use, animal names are capitalized to avoid confusion, thus the Striped Hyena *Hyaena hyaena* can be differentiated from a hyena that happens to be striped.

Cetaceans

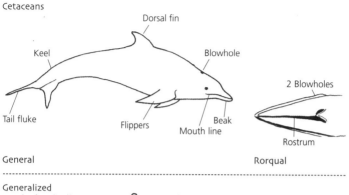

Dorsal fin

Keel

Blowhole

2 Blowholes

Tail fluke

Flippers

Beak

Mouth line

Rostrum

General   Rorqual

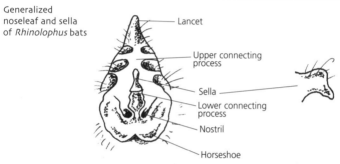

Generalized
noseleaf and sella
of *Rhinolophus* bats

Lancet

Upper connecting
process

Sella

Lower connecting
process

Nostril

Horseshoe

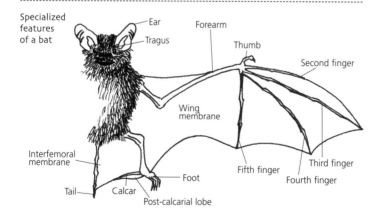

Specialized
features
of a bat

Ear

Tragus

Forearm

Thumb

Second finger

Wing
membrane

Interfemoral
membrane

Foot

Fifth finger

Third finger

Fourth finger

Tail

Calcar

Post-calcarial lobe

Names for parts of mammals used in the text

Identification: The total length of the species is given from the snout to the tip of the tail, excluding any terminal tuft. The tail length is then given. Head and body length, a frequently used parameter, can be calculated by subtracting tail length from total length. Frequently a range is given, reflecting variation in size between individuals, sexes, or subspecies, and it is sometimes surprising how great this variation can be. In the larger mammals, shoulder height is also given and, in the bats, the forearm measurement is given. Where possible, the weight is given. All measurements given are from Egyptian specimens, where possible, or at least from specimens from the region and refer to the published data in Osborn and Helmy (1980), and Harrison and Bates (1991). In many cases, these measurements have been checked to museum specimens (see acknowledgments). Overall appearance and key distinctive features are then given, followed by a detailed description of the external appearance of the species. Where relevant, there may be a discussion of variation, either clinal or by subspecies.

Range and Status: The global range of the species is given followed by a detailed description of the range in Egypt. The status of the species is given based on published records, recent sightings, as well as whether the species is protected under Egyptian law (or has been recorded from any Protected Area) or whether it is covered under international treaties, such as CITES, or listings, such as IUCN.

Habitat: A description is given of the habitat type in which the species is normally found.

Habits: A description of the habits of the species is given. This can include such details as when active (nocturnal, diurnal, or crepuscular), nature of den or burrow, food, acuity of the senses, recorded predators, distinctive behavior, reproductive details including gestation period, breeding season, and litter size. Where possible, these details are based on published records from Egypt but for species very rare in Egypt, e.g., Aardwolf *Proteles cristatus* or Leopard *Panthera pardus*, behavioral details from outside Egypt are given prefaced by the word 'elsewhere' to show that this is the case.

Associated Species: Many mammals, particularly the rodents, can be found in similar habitat, even sharing burrows with other species. When this has been recorded, associated species are given.

Notes: Any useful additional information, particularly of a taxonomic nature, is given under this heading. As stated above, this guide broadly follows the taxonomy of Nowak (1999), but where there are differences these have been noted. Differences with other important works may also be noted such as the taxonomic treatment of the hyraxes in Kingdon (1997) or the Small-spotted Genet *Genetta genetta* in Harrison and Bates (1991).

Similar Species: The key identification characteristics with species likely to give confusion are highlighted in this section.

The Maps

For every species described, there is a map showing the range in which the mammal concerned is likely to be found. These maps are based on published records augmented by reliable field observations.

 Gray shading indicates the area in which the species has definitely been recorded. The species may be expected in suitable habitat within this range according to the mammal's status as given in the text.

 Black stippling indicates a historical range but one in which the species can no longer be expected, or a range in which the animal may now be found only in widely scattered pockets due, for example, to hunting, as in the Nubian Ibex.

 A black dot indicates an isolated record outside the normal range of the species, or a number of records from a very limited area, for example, one of the Western Desert oases.

 A question mark indicates a record that has not been substantiated, is questionable for some reason, or that is very old and for which there have been no more recent sightings.

The Plates

Every species of mammal that has been reliably recorded in Egypt has been illustrated, most groups in color but the bats and the cetaceans in black and white. Each plate shows each individual mammal to scale. The plates are based on three main sources. Where possible, the living mammal has been seen and sketched and the institutions that have made this possible are recognized in the acknowledgments. For the detailed work, particularly with the difficult groups such as the *Gerbillus* gerbils and the *Meriones* jirds, use has been made of well-preserved museum specimens from respected institutions. This has been invaluable and the collections used have likewise been acknowledged. Finally, photographic references have been used. Where necessary or helpful, ink drawings supplement the main illustration, for instance, in the details of the horseshoe bats' noseleaves and sellae, or to show some aspect of behavior.

Measurements

The measurements recorded in the individual species accounts are, for the most part, those from Osborn and Helmy (1980) and in many cases confirmed to museum specimens. These relate to specimens from Egypt as opposed to specimens from outside the country, which may or may not be locally accurate. The measurements for the bats are generally those published by Qumsiyeh (1985). Where there are no Egyptian specimens readily available, as in the Marbled Polecat *Vormela peregusna* for instance, the measurements are taken from the closest reliable source, e.g., Harrison and Bates (1991). Key measurements are as follows.

Total length: Length taken from the tip of the nose to the tip of the tail, excluding any terminal tuft.

Tail length: Length taken from the tip of the tail, excluding any terminal tuft, to the base of the tail.

Head and body length: Length from the base of the tail to the tip of the nose, which may be obtained by subtracting the tail length from the total length.

Ear length: Length of the ear from the tip, excluding any tuft, to the base.

Forearm length: For the bats, the forearm length is given being the distance from the elbow to the wrist, i.e., the ulna.

Hind foot length: Length of the hind foot from the back of the heel to the tip of the toe, excluding the claws.

Shoulder height: The height at the shoulder used in the larger mammals, e.g., the ungulates.

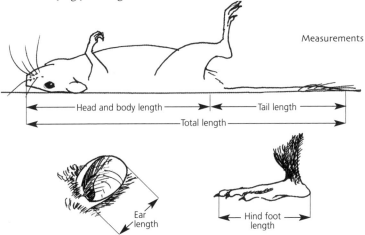

Measurements

Head and body length — Tail length

Total length

Ear length

Hind foot length

All the above measurements are in millimeters for the smaller species, centimeters for the larger species such as the ungulates, or, in the case of the cetaceans, in meters. While this may seem ungainly, this reflects the accuracy with which mammals of various types are generally recorded. To describe Savi's Pygmy Shrew as .07m total length or the Fin Whale as 22,000mm long would be to overgeneralize the former and be overly specific about the latter and to misrepresent them both.

Weight: The weight of a mammal is given, where known, in grams for the smaller species and kilograms for the larger species and tons for the larger cetaceans. See the above note on measurements.

The Insectivores—Order Insectivora

The insectivores is a heterogeneous order of mammals that includes the familiar hedgehogs, shrews, and moles along with the much less familiar golden moles, solenodons, otter shrews, and tenrecs. In the past, elephant shrews, tree shrews, and flying lemurs, or colugos, have also been lumped into the order, though each is now generally assigned to the orders Macroscelidea, Scandentia, and Dermoptera, respectively. Being such a catch-all group, it is difficult to assign general characteristics to the insectivores. However, they are all small, none heavier than 1,500g and generally much smaller and, indeed, include Savi's Pygmy Shrew *Suncus etruscus*, the smallest terrestrial mammal in the world, tipping the scales at a mere 1.5–2.0g. The snout is generally slender, elongated, and highly mobile, the eyes are small or even absent externally, and the fur is short and dense. Many are nocturnal though some, such as the shrews, are active day and night. While the larger insectivores take small vertebrate prey, and some include fruit in their diet, they feed predominantly (as their name suggests) on insects and other invertebrates. Scent is the primary sense used in locating prey.

Everything points to the insectivores being a very ancient group. They have relatively small brains with few convolutions, the testes do not descend into a scrotal sac, the teeth are primitive and well differentiated, and they possess a cloaca, a common exit for the urino-genital and fecal systems. Being an old group though, some members have developed very specialized adaptations: the moles and golden moles have adapted to life underground, the hedgehog and some tenrecs have developed spines, and some shrews, poisonous saliva.

Egypt has relatively few insectivore species, two hedgehogs (though this is questioned on taxonomic grounds) and six shrews, though again, their taxonomy is still a matter of debate. Two of these species, the House Shrew *Suncus murinus* and the Savi's Pygmy Shrew, are known only from single records and certainly the former is introduced. The paucity of insectivores in Egypt is probably due to the lack of suitable habitat. Many insectivores prefer moist habitats where food and water are plentiful (shrews may eat more than their own body weight daily). Relatively few species are found in the desert, hedgehogs being the main exception, though even they are absent from true desert. The Sahara seems to have provided a very effective barrier to the insectivores as only one species, the Ethiopian Hedgehog *Paraechinus aethiopicus*, has been recorded in the south of the country. Of the six shrews, the three *Crocidura* shrews are African species and a fourth, Flower's Shrew *Crocidura floweri*, is endemic, while the others have wide ranges over Eurasia. It is probable that these four current *Crocidura* shrews became isolated in northern Egypt during the climatic warming around six thousand years ago.

The Hedgehogs and Moonrats—Family Erinaceidae
17 species worldwide with 2 in Egypt.

Of the family Erinaceidae, only the hedgehogs, subfamily Erinaceinae, are represented in Egypt, the moonrats being confined to Southeast Asia north to China. The hedgehogs are amongst the most familiar and distinctive insectivores characterized by their relatively large eyes and ears, short, stocky build, and by their entire upper parts being covered in a coat of short spines. Some, but not all, hedgehogs can roll up into a near impregnable ball of spines, the vulnerable underparts and facial region being completely protected.

There is little agreement about the taxonomy of the hedgehogs, even within Egypt. Here, two species are recognized. The Long-eared Hedgehog *Hemiechinus auritus* is quite distinctive and placed in the genus *Hemiechinus* based on cranial characteristics, ear size, and the absence of a spineless 'parting' on the crown. It is probably the most likely of all of Egypt's insectivores to be encountered and certainly the most widespread. The second species, the Ethiopian Hedgehog, is assigned to the genus *Paraechinus*. Some authors, notably Osborn and Helmy (1980), divide the Ethiopian Hedgehog into three species, the Desert, South Sinai, and Ethiopian Hedgehog. The three are differentiated primarily by spine coloration and facial pattern and are considered by most current authors as one species, the Ethiopian Hedgehog *Paraechinus aethiopicus*, with three geographically distinct populations in Egypt: the Mediterranean coastal desert, the mountains of South Sinai, and the southern Eastern Desert, respectively. Because of the degree of variation within hedgehog species (some authors split the European Hedgehog *Erinaecus europaeus* into as many as twelve different species,) it is probable that these three are indeed three isolated populations of the same species, which is how they are treated here.

LONG-EARED HEDGEHOG *Hemiechinus auritus* (Gmelin, 1770)
Pl. 1

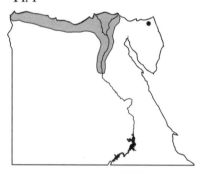

Long-eared Hedgehog
(Hemiechinus auritus)

Subspecies occurring in Egypt: *H. a. aegyptius* and *H. a. libycus*. **Arabic:** *Qunfid tawil al-udhun* **Identification:** Length 151–245mm; Tail 15–39mm; Weight 400–500g. A small, rather pale hedgehog with distinctively large ears. Upper parts covered in rather short spines, reaching some 1.5cm in length along the back, that are brownish at the base with a broad, cream subterminal band and white tip. *H. a. libycus* averages paler than *H. a. aegyptius*, with a slightly shorter tail. The underparts are white, often tinged yellow, and the legs and feet, white to whitish. Rather long-legged, especially apparent when trotting. Facial region without spines, with pale brown fur. Snout long and pointed. Ears large, rounded, whitish, and translucent. Unlike other hedgehogs, they emerge well clear of the spines. Tail short. Wide range of vocalizations including snuffling, growling when threatened, and a cat-like hissing.

Range and status: Wide range from Libya and Egypt across Sinai to Israel, north to Syria, south to eastern Arabia, and east through Iraq, Iran, Afghanistan, Pakistan and India north to Russia and on to Mongolia. In Egypt, restricted to the north. *H. a. aegyptius* found in North Sinai (few records from al-Arish area), much of the Delta, and both sides of the Nile Valley from the barrages south to just south of Beni Suef. Also the Fayoum, where reportedly common. *H. a. libycus* is found on western margin of Delta including Wadi Natrun and across the northern coastal desert to Sallum, seemingly a frequent road kill.

Habitat: Not a species of true desert. Generally found around human settlements and agricultural activity. Also gardens, buildings, and more densely vegetated areas of coastal desert, including salt marshes.

Habits: Nocturnal though occasionally active by day. Spends day in a simple burrow up to 1m long, which it excavates itself. Diet omnivorous, probably largely insects, but will also take fruit and small vertebrates. Possibly an important pest control. Hearing and scent acute. Predators include the Eagle Owl *Bubo bubo* and probably also the Jackal *Canis aureus*, Swamp Cat *Felis chaus*, etc. Unlike other hedgehogs, does not roll into a ball when threatened. Gestation 36–37 days. Female enlarges end of tunnel to make a nest chamber. Elsewhere, breeding season from May to October with peak in summer. Litter size 1–5.

Associated Species: The Long-eared Hedgehog has been found in the burrows of the Fat Sand Rat *Psammomys obesus*.

Notes: Kingdon (1997) also assigns the Ethiopian Hedgehog to the genus *Hemiechinus* but note the differences outlined below in the description of that species.

Similar species: Readily told from the Ethiopian Hedgehog by small size, the very prominent ears, and lack of 'parting' in frontal spines. In hand, also differentiated by uniform pale underside. Habitat is also important—a hedgehog in true desert is not likely to be this species. Porcupines much larger.

ETHIOPIAN HEDGEHOG *Paraechinus aethiopicus* (Ehrenberg, 1833)
Pl. 1

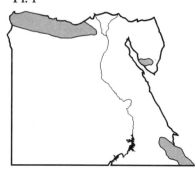

Ethiopian Hedgehog
(Paraechinus aethiopicus)

Subspecies occurring in Egypt: *P. a. aethiopicus*, *P. a. deserti*, and *P. a. dorsalis*.

Arabic: *Qunfid habashi*

Identification: Length 184–258mm; Tail 13–30mm; Weight to 500g. Typical hedgehog form with rather large ears, though not as proportionately large as in the Long-eared Hedgehog. Upper parts covered in spines up to 2.7cm long. Dorsal spines dark, tipped with pale brown. This gives a very dark impression of the back when seen from above, much darker than the Long-eared Hedgehog. Frontal spines divided by a bare patch or parting, extending some 3cm back, but not a field feature. Spines along flank shorter and pale tipped. Underparts white with dark patches, rather variable. Head with rather pointed snout, but not as pointed as Long-eared Hedgehog. Face bicolored or all dark. Snout, chin, and throat dark brown. Forehead to just above eye and down sides pale. Ears large, broad-based, and extending beyond spines. Legs dark brown. Tail short.

Range and status: Egypt, including Sinai, south through Sudan to Somalia and west to Libya, Algeria, Morocco, and Mauritania. Also Arabia north to Lebanon. In Egypt, *P. a. aethiopicus* is known only from the southern Eastern Desert including the Gebel Elba region. *P. a. deserti* recorded from the north coast from Sallum to the western margin of the Delta. *P. a. dorsalis* restricted to South Sinai.

Habitat: Deserts and semi-deserts, rocky wadis, and plains. Also gardens and oases.

Habits: In Egypt, virtually unrecorded and little known elsewhere. Nocturnal, probably most active at dawn and dusk. Spends day in burrow excavated by itself often under dense shrubbery. Home range probably small, within 200–300m of the burrow. Diet presumably

similar to Long-eared Hedgehog, but reportedly more carnivorous including insects, grubs, small vertebrates, and probably fruit, roots, etc. Food may be stored underground. Predators unknown but possibly Eagle Owl. Gestation 30–40 days. Female gives birth to young in burrow or in vegetation. Litter size 1–4. One or more litter per year.
Notes: See comments on family taxonomy.
Similar species: For Long-eared Hedgehog, see previous species.

The Shrews—Family Soricidae
c. 246 species worldwide with 6 in Egypt.

The shrews include some of the world's smallest mammals and indeed worldwide none exceeds 29cm in length or 40g in weight, and most are much smaller. Shrews are characterized not only by their small size but also by their generally mouse-like appearance, the elongated, pointed snout, small eyes and ears, and soft pelage often punctuated by longer bristle hairs, especially on the tail. The absence, presence, and degree to which these occur can be important in identification.

As a consequence of their small size, shrews have a very high surface area to volume ratio and must consume enormous amounts of food relative to their body size just to keep their metabolism going. Thus, shrews are almost constantly on the search for food, which consists largely of insects, arachnids, earthworms, and other invertebrates as well as occasional small vertebrates. In some species, killing is assisted by poisonous saliva. A further adaptation to the shrew's frantic pace of life is refection noted in many, if not all shrew species. In refection, the rectum is extended and licked possibly so that the shrew can obtain nutrients from its food that might otherwise be lost in the feces. As is usual for animals with such a high metabolism, the life span of a shrew is generally very short, few survive beyond their first year. Their short lives are also due to the fact that shrews are born with one set of teeth. Once these wear out then the animal will die of starvation. They are thought to suffer less from predation than, for example, the smaller rodents, due to their distasteful flesh caused by noxious secretions from skin glands. That being said, their remains have been found in the pellets of Barn Owls *Tyto alba*.

The paucity of shrew species, mentioned under the general discussion of the insectivores, is paralleled by there being relatively few specimens of most species taken in Egypt. Indeed, only the Greater Musk Shrew *Crocidura flavescens* can be called at all widespread or common, and no other shrew has been recorded south of the Fayoum. This may reflect the genuine scarcity of shrews in Egypt, though the Nile Delta and Valley would appear to provide the moister habitat, rich in food and water, that shrews favor. It more likely reflects the fact that they are easily overlooked. Their small size (combined with preference for densely vegetated habitat) and the difficulty in trapping them makes them hard to locate or study. Savi's Pygmy Shrew is too small to trap in conventional traps even if it were attracted by the same bait put out for the more vegetarian rodents. While it is probable that the isolated record of the House Shrew is of a single, introduced animal at the port of Suez, the Delta may support larger numbers of Savi's Pygmy Shrew than the number of specimens collected would seem to indicate. The records of mummified specimens of Flower's Shrew and the Dwarf Shrew *Crocidura nana* from Thebes in Upper Egypt may indicate that shrews did have a wider range in Egypt in pharaonic times when the climate was milder. Mummified shrews of as yet unidentified species have also been found at Saqqara.

It is probable, depending on their numbers and population density, that shrews consume a large number of agricultural insect pests.

Shrews are often referred to in Arabic by the generic terms for mouse or weasel: *far, 'irsa.*

GREATER MUSK SHREW (GIANT MUSK SHREW, AFRICAN GIANT SHREW, GIANT SHREW) *Crocidura flavescens* (I. Geoffroy St-Hilaire, 1827)
Pl. 2
Subspecies occurring in Egypt: *C. f. deltae.*
Arabic: *Zibab 'imlaq*
Identification: Length 163–219mm; Tail 57–84mm; Weight c. 40g. Large, rather dark, long-tailed shrew. Upper parts uniform dark brown, glossy on flanks, fur soft, short, and dense. Underparts dark grayish. Feet either grayish or brown. Head with typical pointed

snout. Teeth white. Eyes visible but small. Ears large, sparsely haired, and, despite size, barely stand out from dense fur. Tail dark brown, about half of head and body length, sparse hair with long bristles except along last third of length.

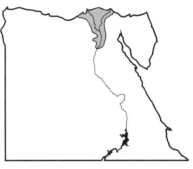

Greater Musk Shrew
(*Crocidura flavescens*)

Range and status: Patchily throughout Africa from Cape along eastern southern Africa north through East Africa to Ethiopia, Sudan, and Egypt, and west to Sierra Leone. In Egypt, easily the most widespread shrew, but even so, confined to the Delta and the Nile Valley south to Dahshur. Historically abundant in the region south of Damietta. Reported common at Abu Rawash and around Zaqaziq. Also the Fayoum where reportedly common.

Habitat: Moist areas with dense vegetation including canal embankments, irrigated fields, and cultivated areas. In the Fayoum, around settlements and gardens.

Habits: Like most shrews, may be active throughout the day and night. Territorial and solitary. Builds a nest of damp, matted grass. Diet mainly insects and other invertebrates including snails, but may also take small vertebrates with frog carcasses having been recorded in their nests. Predators include the Eagle Owl. Gestation c. 18 days. Litter size 2–6. Elsewhere, breeding season during wetter months.

Similar species: Other shrews. Distinguished by much larger size except from the very rare House Shrew, which has a proportionately shorter tail, thick at base, and lighter underside. Distinguished from the Flower's Shrew, Lesser White-toothed Shrew *Crocidura suaveolens*, and Dwarf Shrew by dark underside and uniformly dark tail. All much smaller. In hand, first has tail bristles only along first half of tail; other two have bristles along entire length.

FLOWER'S SHREW *Crocidura floweri* Dollman, 1915
Pl. 2

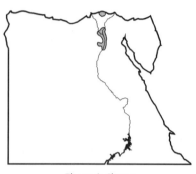

Flower's Shrew
(*Crocidura floweri*)

Monotypic
Identification: Length 112–129mm; Tail 55–58mm. Small shrew, endemic to Egypt. Upper parts rather pale pinkish brown. Flanks and underparts whitish. Feet grayish white. Tail proportionately long, up to 75% of the head and body length, reddish brown above, whitish below with bristles only on first half. Bristles sparse and grayish white.

Range and status: Only known from Egypt where confined to Nile Delta and the Fayoum. Specimens recorded from Giza, Baltim, and a single record from the Fayoum. Very rare with no recent published records. Possibly extinct.

Habitat: Agricultural areas.

Habits: Unknown. Presumably similar to other shrews. Predators include Cattle Egret *Bubulcus ibis* as one specimen recorded from the stomach of that species.

Similar species: Other small shrews from which distinguished by cinnamon tinge to upper parts and sparse bristles only extending along basal half of the tail.

DWARF SHREW (MUSK SHREW) *Crocidura nana* Dobson, 1890
Pl. 2

Monotypic

Arabic: *Zibab qazam*

Identification: Length 76–102mm; Tail 28–40mm. Very small, pale, long-tailed shrew. Coat short and dense, upper parts grayish, tinged brown, flanks paler and pale gray below. Head typically shrew-like, grayish with whitish throat and chin. Feet whitish, almost naked. Tail proportionately long, about 60% of head and body length. Gray above, whitish below with numerous, whitish bristles along entire length.

Range and status: From Egypt south through East Africa to Zimbabwe. In Egypt, recorded only from the southernmost part of the Delta south to Cairo (1 record).

Dwarf Shrew
(*Crocidura nana*)

Habitat: Similar to other shrews. Recorded from moist farmland, canal banks, etc.

Habits: Unknown. Presumably similar to other shrews. Nest reportedly built of twigs and cotton bolls.

Notes: A further species, *Crocidura religiosa*, has been described from ancient Egyptian mummies (the Dwarf Shrew itself is known from mummified remains from Thebes). However, many authors reject this species and, unless a living specimen is found, it cannot be distinguished in the field and can only be identified from skeletal characteristics; thus, it has not been included. Bonhote (1909) claimed to have obtained one live specimen of *C. religiosa* at Abu Rawash, but provided no details.

Similar species: Other small shrews. For Flower's Shrew, see previous species. From Lesser White-toothed Shrew told by proportionately longer tail and paler, grayer color above. Species' known distributions in Egypt do not remotely overlap. Distinguished from Savi's Pygmy Shrew by larger size and proportionately longer tail.

LESSER WHITE-TOOTHED SHREW *Crocidura suaveolens* (Pallas, 1811) **Pl. 2**

Subspecies occurring in Egypt: *C. s. portali* and *C. s. matruhensis*.

Identification: Length 80–112mm; Tail 25–40mm; Weight 3.5g. Small shrew with proportionately short tail. Upper parts brown to brownish gray, underparts whitish with no sharp demarcation along flanks. Feet whitish. Head with broad snout narrowing quickly to elongated proboscis. Ears rather small but distinct, standing out from fur. Eyes typi-

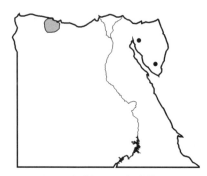

Lesser White-toothed Shrew
(Crocidura suaveolens)

cally small. Tail same color as upper parts, indistinctly paler below. Bristles extend entire length of tail. Has a distinct musky odor.

Range and status: Very wide range over much of southern Europe, west to Scilly Isles and north to northern Poland, east throughout Asia from Middle East to China, Korea to Japan. Isolated records from Arabia. In Egypt, very few records. *C. s. portali* recorded from South Sinai west of St. Katherine's Monastery, and from near Suez. *C. s. matruhensis* is known from only a few specimens taken west of Mersa Matruh. Despite the paucity of Egyptian records, the wide range of this species elsewhere may mean that it could turn up in areas in Egypt other than those in which it has so far been recorded.

Habitat: In Sinai, taken from 1,500m inside a small monastery west of St. Katherine. On the north coast, taken from Fat Sand Rat burrows in salty depressions near coast. Elsewhere, recorded from a wide range of habitats including woodland, gardens, marshlands, rocky hill slopes, vegetated dunes, and coastal plains.

Habits: Unknown in Egypt. Elsewhere, little known in wild but probably much like other shrews. Active throughout day and night though with peak in evening. Does not excavate a burrow but lives in cavities or crevices or in thick tangles of vegetation. Diet consists of insects, snails, worms, and other invertebrates. Gestation 24–32 days. Litter size 1–7. Once mobile, young follow female around by caravaning, where first of litter grabs mother by the rump, the second grabs the first, etc., forming a train.

Note: Certain authors reorganize *C. s. matruhensis* as a separate species, *C. whitakeri.*

Similar species: Other small shrews. For Flower's Shrew and Dwarf Shrew, see those species. Much smaller than House Shrew. Much larger than Savi's Pygmy Shrew with relatively smaller ears.

HOUSE SHREW *Suncus murinus* (Linnaeus, 1766)
Pl. 2

Subspecies occurring in Egypt: tentatively *S. m. sacer.*
Identification: Length 155–240mm; Tail 50–85mm. A large, robust shrew with a proportionately rather short tail. Coat short. Upper parts brown, hairs gray at base, underparts grayish white with indistinct demarcation. Feet white. Head large with relatively small, rounded ears that still project beyond fur. Tail

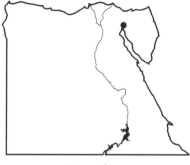

House Shrew
(Suncus murinus)

about one half of head and body length. Very thick at base narrowing toward tip. Brown with silvery bristles along entire length.
Range and status: Largely Asiatic, from New Guinea west through Southeast Asia to India and Sri Lanka and north to Taiwan and Japan. Outside this range also in isolated populations at seaports throughout Arabia to Egypt and Sudan. In Egypt there is one (two according to Wassif and Hoogstral [1953]) record of the species at Suez. This was almost certainly ship-borne and there is no evidence that the House Shrew has established a viable population at the port.
Habitat: The House Shrew appears to be a commensal, at least in the Arabian region, found in houses, warehouses, and other buildings. Also recorded from garbage heaps, gardens, and walls.
Habits: Unknown in Egypt. Probably as other shrews but much more tied to humans. Elsewhere, reported to be largely nocturnal and noisy. Recorded from Barn Owl pellets.
Similar species: In Egypt, only the Greater Musk Shrew is as large. House Shrew can be distinguished by proportionately shorter tail with much thicker stock and silvery bristles along entire length. Range and habitat different.

SAVI'S PYGMY SHREW (PYGMY WHITE-TOOTHED SHREW, COMMON
DWARF SHREW, SAVI'S DWARF SHREW, ETRUSCAN SHREW) *Suncus
etruscus* (Savi, 1822)

Pl. 2

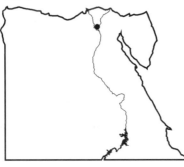

Savi's Pygmy Shrew
(Suncus etruscus)

Single Egyptian specimen probably *S. e. etruscus.*

Identification: Length 62–81mm; Tail 21–32mm; Weight 1.5–2.0g. A tiny shrew, one of the smallest mammals in the world, with relatively large ears. Upper parts gray-brown with reddish tinge, long hairs interspersed in short dense coat. Underparts grayish, hairs gray, tipped white, with indistinct demarcation. Feet whitish. Ears large and stand well clear of fur. Tail narrow, over half head and body length, brown above, paler below, and scattered with bristles along entire length.

Range and status: Much of southern Europe, including the Canary Isles, from Iberia east to Turkey and on to Iraq, Iran, India to China and parts of Southeast Asia to Borneo. Also North Africa south to Ethiopia and Madagascar. In Egypt, known only from one specimen in the Paris Museum taken in the Delta. It is not known whether this shrew is just very rare, possibly extinct, or whether it is so small it is overlooked. Apparently it is too small to be caught in conventional mousetraps.

Habitat: Unknown in Egypt. Elsewhere, from farmland, in gardens, olive groves, along old walls, and buildings.

Habits: Little known but presumably much as other shrews. Nests in cavities in walls or rocks, or beneath tree roots.

Similar species: See other small shrews. Minute size and relatively large ears should be diagnostic.

The Bats—Order Chiroptera

The bats are unique amongst the mammals in their ability to fly. While other mammals, such as the flying squirrels, the scaly-tails, and the peculiar flying lemurs—none of which occur in Egypt—can glide on skin membranes between their outstretched limbs, it is only the bats that are capable of true, flapping flight. In order to do this, the bats have evolved a body plan radically different from any other group of mammals. The forelimbs have essentially become the wings with the fingers greatly elongated to form a framework over which the flight membrane (a slender membrane of skin and narrow layers of muscle fibers) is stretched. The membrane extends back to the much-reduced hind limbs and, in many species, to the tail as well. The extent and shape of the flight membrane is important in identification.

In devoting the forelimbs to flight, the bats have largely sacrificed their ability to move on the ground, although, amongst the Egyptian species, the free-tailed bats can move relatively rapidly across walls, etc., scuttling along using their folded wings as legs. Most bats roost upside down, suspended by their hind feet or clinging by their hind feet from a cave, tomb, or building wall. At rest, the wings may be wrapped around the bat, as in the horseshoe bats, or folded up on either side as in the rat-tailed bats.

Most bats are nocturnal, though the actual time of flight varies, some species emerging at dusk, others feeding later at night.

The Old World Fruit Bats—Suborder Megachiroptera
Family Pteropodidae
Approx. 173 species with 1 in Egypt

The fruit bats are the largest bats and differ from other bats, not only in size but also in their prominent eyes, relatively small ears with no tragus, two wing claws, and the absence, or great reduction, of the tail and interfemoral (tail) membrane. There is only one Egyptian species, the Egyptian Fruit Bat *Rousettus egyptiacus*.

EGYPTIAN FRUIT BAT *Rousettus egyptiacus* (E. Geoffroy St-Hilaire, 1810)
Pl. 3

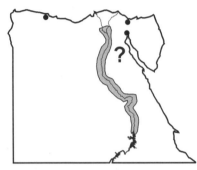

Egyptian Fruit Bat
(Rousettus egyptiacus)

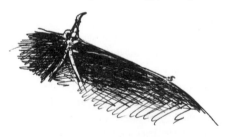

Subspecies occurring in Egypt: *R. e. egyptiacus*.
Arabic: *Khuffash al-fak-ha misri*
Identification: Length 126–167mm; Tail 8–19mm; Forearm 84–99mm; Weight 130g. Male probably slightly larger than female. Large, short-tailed bat with fox-like face and prominent eyes. With a wingspan of up to 60cm, this is by far the largest Egyptian bat, identifiable by size alone, but in the hand, the second finger is clawed, tragus absent, and ear margin complete. Overall pale grayish to brown in color, slightly darker above. Male has elongated hair on the throat related to scent glands. Inter-femoral membrane much reduced and tail very short (appears tailless in flight).

Range and status: Eastern Mediterranean, including southern Turkey and Cyprus, east to Iran and Pakistan. Arabia and much of Africa. In Egypt, found throughout the Nile Valley south to Aswan. Also in cities including Cairo (inc. Mohandiseen, Giza, Zamalek, and Garden City) where common. Other records from north coast near Mersa Matruh, Port Said, and Suez.

Habitat: Cultivated areas to desert margins, towns, and cities. Preferred roosts are mosques, deserted buildings, tombs, monuments (inc. Giza pyramids), etc., that are generally humid with some indirect light.

Habits: Nocturnal with peaks at dusk and dawn. Often in large colonies of hundreds, even thousands. Although they have large eyes, these fruit bats can also echolocate and, thus, fly in complete darkness. Probably use sonar to find their way out of their roost and then fly by sight. Voice audible, metallic squeak emitted by clicking the tongue. Feeds mainly on fruit, e.g., mango, date, figs, etc., and leaves. Also seen at *Bombax* flowers although not proven to feed on them. Can be found by directing torchlight at fruiting tree and seeing orange-yellow eye reflection. Gestation 4 months, breeding February/March to May in Egypt, though elsewhere in region breeds twice a year. One occasionally 2 young. Predators include large falcons and owls, especially the Barn Owl *Tyto alba.*

Notes: Unpopular as they can deface monuments and eat fruit crops. However, research in Israel showed that Egyptian Fruit Bats ate only ripe fruit rather than fruit for packing, processing, and transport, which is picked green. Moreover, gassing of roosts resulted in decimation of insect-eating bats and a rise in the level of insect pests. In Africa, they have been shown to be important pollinators and seed dispersers.

Similar species: In Egypt, only *Tadarida* bats reach anything near the size of Egyptian Fruit Bats and can be further distinguished by much larger ears, darker color, and prominent tail.

The Insectivorous Bats—Suborder Microchiroptera

The insectivorous bats form a much larger group than the fruit bats and are represented in Egypt by twenty-one different species, although several are known only from very few records. Unlike the fruit bats, the insectivorous bats hunt actively for insects, which they find using echolocation, a form of sonar. The face of most bats may appear grotesque to human eyes, but it is actually adapted to emit and receive high frequency or ultra-sonic sound waves. The bat releases a constant series of clicks through the mouth or nostrils that are then reflected off any obstacle or prey. The rebounded sound is picked up by the ears that are often very large and possess a lobe known as the tragus (absent in the horseshoe bats) at the opening. The rapidity and sensitivity of the bats' echolocation system can be seen by simply watching bats flying around at speed at dusk, rapidly dodging and weaving in pursuit of flying insects. The insect-eating bats are probably important agents in insect pest control in agricultural areas. Since the insectivorous bats hunt by echolocation, the eyes in many species have been reduced to a greater or lesser extent.

Note: All bat tragi and noseleaves after Qumsiyah (1985) and Harrison and Bates (1991).

Identification in the field of the twenty-one species of insectivorous bats is very difficult, and in trying to identify free-flying bats, the type of flight is important: high, low, over water, fast, slow, etc., as is the habitat, and the geographical location. For instance, a large-eared bat seen in arid desert is far more likely to be Hemprich's Long-eared Bat *Otonycteris hemprichii* than the Gray Long-eared Bat *Plecotus austriacus* or the Egyptian Slit-faced Bat *Nycteris thebaica*. In the hand, identification to family level can be made by looking at ear and nose patterns, tail and interfemoral membrane form, wing shape, color, and bare areas. These will all be discussed under individual families.

Rat-tailed Bats—Family Rhinopomatidae
3 species worldwide with 2 in Egypt.

Distinguished from all other Egyptian bats by the greatly elongated and very slender tail and virtual absence of interfemoral membrane. In practice, this feature is impossible to see in flight and is so slender that it can only just be made out when the bat is at rest. Rat-tailed bats do not hang from their roosts but cling upside down to the roost wall. The tail is held slightly curved out over the bat. At closer quarters, further distinguished by the pig-like snout, ears meeting over the forehead, and simple tragus. Other than size, the two Egyptian species are very difficult to tell apart, the critical difference being the relative lengths of the tail and the forearm, the tail being shorter than the forearm in the Larger Rat-tailed Bat *Rhinopoma microphyllum* and longer in the Lesser Rat-tailed Bat *Rhinopoma hardwickii*. The Larger is a much rarer species.

LARGER RAT-TAILED BAT (GREATER MOUSE-TAILED BAT) *Rhinopoma microphyllum* (Brunnich, 1782)

Pl. 4

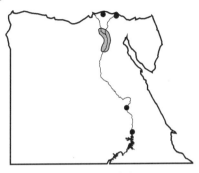

Larger Rat-tailed Bat
(Rhinopoma microphyllum)

Subspecies occurring in Egypt:
R. m. microphyllum.

Arabic: *Abu dhayl al-kabir*

Identification: Length 124–146mm; Tail 52–65mm; Forearm 64–71mm. Rat-tailed bats distinguished by slightly upturned snout and long, slender tail though this feature is almost impossible to see in flight. At rest, e.g., on cave wall, tail curves forward in crescent. Largest of the rat-tailed bats, though size is not a useful field feature. Eyes distinct and well-developed ears with sickle-shaped tragus. Thumbs elongated and feet slender. Long tail surrounded by flight membrane only at base. Fur fine, pale gray-brown above, slightly paler below. Lower back and abdomen, face, lips, and upper throat naked. Specimens from Upper Egypt may relate to *R. m. tropica.*

Range and status: Africa from Morocco south to Nigeria, and east to Sudan and Egypt. Also further east to Arabia, Iran, Pakistan, India, and to Thailand and Sumatra. In Egypt, recorded from Delta, Cairo region, Luxor (mummified), and Aswan (?). Everywhere much rarer than Lesser Rat-tailed Bat.

Habitat: Cultivated areas of Nile Valley and Delta. Has been found in same roosts as Lesser Rat-tailed Bats, though always much rarer.

Habits: Little known. In India, breeding thought to be in June. No evidence from Egypt.

Similar species: See Lesser Rat-tailed Bat below.

LESSER RAT-TAILED BAT (LESSER MOUSE-TAILED BAT) *Rhinopoma hardwickii* Gray, 1831

Pl. 4

Subspecies occurring in Egypt: *R. h. arabium* and *R. h. cystops.*

Arabic: *Abu dhayl al-saghir*

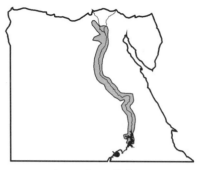

Lesser Rat-tailed Bat
(Rhinopoma hardwickii)

Identification: Length 101–144mm; Tail 46–73mm; Forearm 47–59mm; Weight 10–15g. Very delicately built rat-tailed bat with very long, slender tail only bounded by flight membrane at base. In hand, forearm-to-tail ratio important. Fur fine. Color uniform pale gray-brown, paler below but with some variation. Two subspecies in Egypt. *R.h. arabium* from northern Egypt, inc. Cairo and the Fayoum, browner above and below, though can be gray tinged. *R.h. cystops* from Luxor south to Sudanese border, on average smaller, paler, grayer, described as 'pearl gray' below. Flight peculiar, fluttery, and bird-like.

Range and status: East and North Africa. Arabia east to Iran, Pakistan, India to Indonesia. In Egypt, *R. h. arabium* in Cairo and its environs, inc. Wadi Digla, recorded north to Wadi Natrun, and in the Fayoum (Qasr Qarun). *R. h. cystops* from Asyut region south to Luxor (common in Dendera Temple), and south to Aswan and Sudanese border.

Habitat: Dry caverns and caves, often roosting deep in rock, ruins, temples, tunnels, mosques, and buildings, often on desert margins. Roosts may be shared by much rarer Larger Rat-tailed Bat and with Egyptian Free-tailed Bat *Tadarida aegyptiaca*.

Habits: Emerges at dusk. Colonies fairly small though up to 200 have been recorded. Food probably small insects. In autumn, lays down fat reserves and, thus, can remain active year round. In roosts, voice is audible as high-pitched chirp. Females with suckling young observed just south of Aswan at Kalabsha in July. When sharing with Egyptian Free-tailed Bat, tends to roost higher than latter species and readily distinguished by smaller size, more rapid flight, and higher voice.

Similar species: Larger Rat-tailed Bat is larger, more heavily built and, most importantly, its tail is shorter than forearm (longer in present species).

Sheath-tailed and Tomb Bats—Family Emballonuridae

c. 50 species worldwide with 2 in Egypt.

Small- to medium-sized bats. Egyptian species small, with long ears and variable tragus. Eyes more prominent than in most insectivorous bats. In the hand, the tail is diagnostic, emerging from the interfemoral membrane about two-thirds of the way down its length with the latter third free, but often not exceeding the interfemoral membrane in length. When roosting, clings to walls not hanging free. Egyptian species difficult to tell apart except in the hand where differences are obvious, otherwise habits are best distinction.

GEOFFROY'S TOMB BAT (TOMB BAT) *Taphozous perforatus* (E. Geoffroy St-Hilaire, 1818)

Pl. 4

Subspecies occurring in Egypt: probably *T. p. perforatus*.

Arabic: *Abu buz al-saghir, Khuffash al-maqabir*

Identification: Length 94–112mm; Tail 20–27mm; Forearm 61–66mm. Male slightly smaller than Female. Small bat with narrow-based, almost mushroom-shaped tragus. Ears long and narrow with hair tufts at the base of the back of each ear. Fur silky,

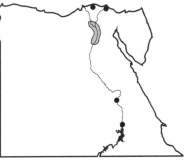

Geoffroy's Tomb Bat
(Taphozous perforatus)

extending to base of tail both above and below. Above, brown with hair bases white; below, variable grayish brown fur extending to tail root above and below. Wing membrane brownish with pale outer edge to forearm. Tail does not exceed interfemoral membrane in length.

Range and status: Africa south of Sahara to Botswana, north to Sudan and Egypt. East to Arabia, Pakistan, and northwestern India. In Egypt, recorded from the Delta, Wadi Natrun,

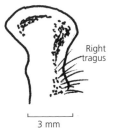

Right tragus

3 mm

Geoffroy's Tomb Bat
Taphozous perforatus

Cairo and environs (inc. Abu Rawash, Saqqara, and Giza), Fayoum, south along Nile Valley to Luxor and Aswan, down to Sudanese border. Red Sea coast near Quseir. Can occur in large numbers.

Habitat: Roosts in deep caverns, limestone caves, sometimes near sea, and crevices, old buildings, and ruins. Degree of light does not seem to be important.

Habits: Poorly known. Roosts can be large but elsewhere reported to be between 6–10 individuals. Often hangs from wall close to ground. Flies at dusk (even recorded flying by day) but also at night. Breeding probably April/May in Egypt. Generally one young.

Similar species: Egyptian Sheath-tailed Bat *Taphozous nudiventris*, see below.

EGYPTIAN SHEATH-TAILED BAT (NAKED-BELLIED TOMB BAT, NAKED-RUMPED BAT) *Taphozous nudiventris* Cretzschmar, 1830

Pl. 4

Egyptian Sheath-tailed Bat
(Taphozous nudiventris)

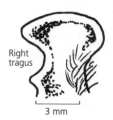

Right tragus

3 mm

Egyptian Sheath-tailed Bat
Taphozous nudiventris

Subspecies occurring in Egypt: *T. n. nudiventris*.

Arabic: *Abu buz al-kabir*

Identification: Length 112–136mm; Tail 22–34mm; Forearm 26–29mm. Large sheath-tailed bat. Ears long, narrow, blunt-tipped, and set wide apart on head. Tragus distinctive, thick-based with clear lobes and flat top. Muzzle, chin, sides of face, and lower back and front are entirely naked. Fur short, pale brown on back, grayer and lighter below. Sexes distinguished in hand by presence of small gular sac in male. Tail length very variable.

Range and status: Africa south to Democratic Republic of Congo. Arabia and southwestern Asia east to Burma. In Egypt, Delta (Gharbiya Governorate), Cairo and environs (inc. Giza where it roosts at each of the three pyramids), Fayoum, Nile Valley south to Luxor (Karnak). Red Sea near Quseir. Can be common.

Habitat: Roosts in very dense colonies in crevices in cliffs, often sandstone, old ruins, wells, mosques, etc.

Habits: Dense colonies can be detected by accumulated droppings and unpleasant 'rubbery' smell. Voice loud squeak, described as metallic. Tends to hang from walls rather than roofs. May travel long distances from roosting to feeding sites. Flight fast, high, and direct. In Egypt, accumulates fat in autumn and flies throughout the year. In the Delta, said to be especially common in July and August when it feeds on the adult moth of the Cotton Leaf Worm *Spodoptera littoralis*, a pest species. Remains have been found in Barn Owl pellets.

Similar species: Geoffroy's Tomb Bat is fully furred with (in the hand) a differently shaped tragus. Egyptian Slit-faced Bat, also common around the pyramids, has a much more erratic flight, is fully furred, and tail is totally surrounded by flight membrane.

Slit-faced Bats—Family Nycteridae
11 species worldwide with 1 in Egypt

Small- to medium-sized insectivorous bats with large, elongated ears that are held straight up from the head, not angled outward. Tragus small. The name comes from the distinct slit or furrow down the center of the face, which runs from between the eyes to the snout. Eyes very small. Tail structure is unique in that the tip is T-shaped.

EGYPTIAN SLIT-FACED BAT *Nycteris thebaica* E. Geoffroy St-Hilaire, 1813
Pl. 4
Subspecies occurring in Egypt: *N. t. thebaica*.
Arabic: *Khuffash tiba*
Identification: Length 84–126mm; Tail 45–56mm; Forearm 43–49mm. Small, broad-winged bat with rather broad, erect ears up to 40mm long. Tragus simple, rounded, and relatively small. Tail long but entirely enclosed within flight membrane and, uniquely amongst Egyptian bats, tip T-shaped. Wings broad and rounded. Distinguished from other Egyptian bats by lobed groove running down center of face. Eyes small but distinct. Fur rather long. Brownish to brown-gray above, paler below. Naked skin of muzzle and base of ears pale pink.

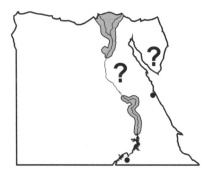

Egyptian Slit-faced Bat
(Nycteris thebaica)

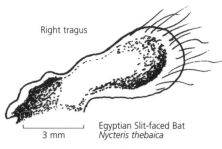

Right tragus

3 mm   Egyptian Slit-faced Bat
        Nycteris thebaica

Range and status: Throughout Africa except Sahara and rainforest regions. Palestine and Israel and western Arabia, vagrant in Corfu. In Egypt, Delta, Wadi Natrun, Cairo and its environs (inc. Giza pyramids), Fayoum (inc. Shakshuk), and Nile Valley south to Dendera, Luxor, Aswan, and Sudanese border. Records from Red Sea at Quseir.

Habitat: Wide habitat tolerance. Roosts in caves, tombs, ruins, houses, wells, etc. Not a desert species.

Habits: Roosts in colonies ranging from a few individuals to several hundred. Reported to fly early and feeds on insects, inc. moths, grasshoppers, beetles, and also, reportedly, scorpions. Prey carried to regular feeding points that can be told by the accumulation of inedible prey parts. Flight erratic. In Egypt, from early March males leave colonies and females form maternity roosts, though individual males remain. Breeding April–July. Generally 1 young. Outside breeding season may roost with Arabian Horseshoe Bat *Rhinolophus clivosus*.

Notes: According to Qumsiyah (1985), a further species *Nycteris hispida* may occur in southern Egypt since there is an unconfirmed record from northern Sudan.

Similar species: See Egyptian Sheathtailed Bat. In the hand, the facial slit and the T-shaped tip to the tail make this species unique in Egypt.

Horseshoe Bats—Family Rhinolophidae

69 species worldwide, 3 in Egypt.

A well-defined family of bats, but identification of individual species—even in the hand—can be very difficult. In the hand, horseshoe bats can be told from all other bats, except the leaf-nosed bats, by the absence of a tragus and by the distinctive horseshoe-shaped nose structure to which they owe their name. The ears are large and broad. Hind limbs are poorly developed. Tail entirely enclosed in interfemoral membrane, though in many species the membrane comes to a point at the tail tip. The wings are relatively broad, making these bats amongst

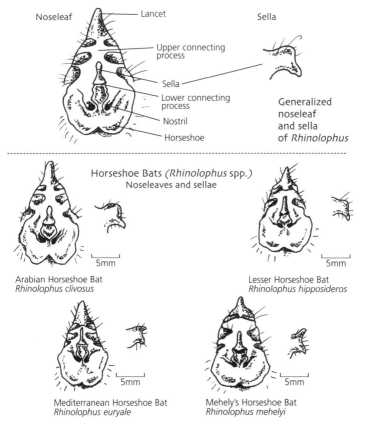

Noseleaf — Lancet — Sella

Upper connecting process

Sella

Lower connecting process

Nostril

Horseshoe

Generalized noseleaf and sella of *Rhinolophus*

Horseshoe Bats *(Rhinolophus* spp.)
Noseleaves and sellae

5mm

Arabian Horseshoe Bat
Rhinolophus clivosus

5mm

Lesser Horseshoe Bat
Rhinolophus hipposideros

5mm

Mediterranean Horseshoe Bat
Rhinolophus euryale

5mm

Mehely's Horseshoe Bat
Rhinolophus mehelyi

the most maneuverable flyers in the order. At rest, horseshoe bats wrap their wings around themselves. They roost hanging, rather than cling-ing, from walls and ceilings. Colonies may be huge, but some species are solitary and at least one species, the Lesser Horseshoe Bat *Rhinolophus hipposideros*, is known only from a single specimen in Egypt.

Horseshoe bats are very difficult to tell apart, even in the hand. Indeed, the identification of one species, Mehely's Horseshoe Bat *Rhinolophus mehelyi*, is still in some doubt due to confusion with the extremely similar Mediterranean Horseshoe Bat *Rhinolophus euryale*. In this instance, the current author has followed Qumsiyah (1985). In the field, key considerations should be location and habi-tat. In the hand, size and detailed study of the noseleaf pattern is essential. For skull characteristics of most species, see Harrison and Bates (1991). At roosts, horseshoe bat colonies tend to be loose and wide-spaced as opposed to the very dense colonies formed by some species of vesper bats. At rest, the *Rhinolophus* bats hang with the wings wrapped tightly around them.

ARABIAN HORSESHOE BAT (GEOFFROY'S HORSESHOE BAT)
Rhinolophus clivosus Cretzschmar, 1828
Pl. 5

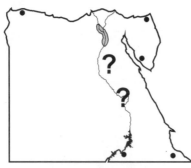

Arabian Horseshoe Bat
(Rhinolophus clivosus)

Subspecies occurring in Egypt: *R. c. clivosus* and *R. c. brachyg-nathus*.

Arabic: *al-Khuffash hadwat al-faras*

Identification: Length 72–82mm; Tail 25–32mm; Fore-arm 48–50mm; Weight 17g. Large horseshoe bat. Connect-ing process of the sella blunt. For nose pattern, see diagram. Nominate sub sp. *R. c. clivosus* color variable from smoky gray in Sinai to dark gray in Libya. Southern populations browner. Underside dull brown-gray. Wing and tail membranes dark brown. *R. c. brachygnathus* smaller and darker colored.

Range and status: Across North Africa, most of East and southern Africa. Western and southwestern Arabia and southwestern Asia. In Egypt, *R. c. clivosus* recorded from North and South Sinai (al-Arish and Wadi Feiran), and Sudan Government Administration Area. *R. c. brachygnathus* recorded from the north coast, Wadi Natrun, Cairo and environs, and south along the Nile Valley. Everywhere scarce.

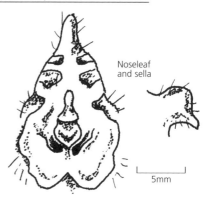

Noseleaf and sella

5mm

Arabian Horseshoe Bat
Rhinolophus clivosus

Habitat: Old storehouses, stone buildings, and caves. A semi-desert and savanna species. Harrison and Bates (1991) note a preference for buildings.

Habits: Little known in Egypt. In Africa, colonies can number several thousand, though in Arabia much smaller roosts recorded. Due to the scarcity of Egyptian records, it would seem that roosts here are small. Females give birth to a single young in summer.

Similar species: Other horseshoe bats. Species identification very difficult unless the bat is in hand, see diagrams of sellae and noseleaves.

LESSER HORSESHOE BAT *Rhinolophus hipposideros* (Bechstein, 1800)
Pl. 5

Subspecies occurring in Egypt: *R. h. minimus*.

Identification: Length 58–67mm; Tail 19–24mm; Forearm 36–38mm (measurements from Jordanian specimens); Weight 3.5–10g. Smallest horseshoe bat of the region and distinguished by delicate form. Sella hook-like, pointing forward and down, unique in region. Ears rela-

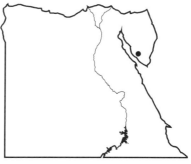

Lesser Horseshoe Bat
(Rhinolophus hipposideros)

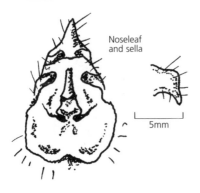

Noseleaf
and sella

5mm

Lesser Horseshoe Bat
Rhinolophus hipposideros

tively large. Fur long and dense. Color rather variable, brown to gray-brown above, paler below.

Range and status: Southern Europe and northwestern Africa. Ethiopia and Sudan. Arabia, southwestern Asia to northwestern India. In Egypt, known only from one specimen taken in Wadi Feiran and assigned to *R. h. minimus*. In light of Afro-asian distribution, could prove more widespread.

Habitat: Roosts in caves, cellars, and buildings. Sole Egyptian specimen found deep inside a cave.

Habits: Usually solitary at roost but females may roost colonially in summer. Elsewhere, flies late and throughout the night. Flight low (c. 2–3m) and fast with very rapid, vibrating, wing beats. Does not follow set routes when foraging. Voice pitched low. Generally 1 young born in summer. Individuals found hibernating in Jordan.

Similar species: Other horseshoe bats. See diagrams of sellae and noseleaves. In hand, small size distinctive.

MEHELY'S HORSESHOE BAT *Rhinolophus mehelyi* Matschie, 1901
Pl. 5

Subspecies occurring in Egypt: *R. m. mehelyi*.

Identification: Length 81–86mm; Tail 21–26mm; Forearm 46–50mm. Medium-sized horseshoe bat with relatively short and broad ears. Sella with sharply pointed upper appendage. Lancet very narrow. Gray-brown to warm brown above, paler below.

Range and status: Southern Europe, North Africa, east to northern Arabia and Iran. In Egypt, recorded from north coast at Alexandria and to the west, and Cairo and Giza, south to Saqqara. Rare in Egypt.

Habitat: Roosts mainly in caves. Elsewhere, colonies of up to 30,000 bats have been recorded, but such records are unlikely in Egypt owing to its scarcity. Although recorded mainly from near urban areas in Egypt, it is considered more of a desert species.

Habits: Little known, especially in Egypt. Litter size estimated at two.

Notes: Mehely's Horseshoe bat is very similar to the Mediterranean Horseshoe Bat *Rhinolophus euryale* and can only safely be distinguished from it by detailed analysis of the noseleaf. The Mediterranean Horseshoe Bat was included on the Egyptian list but Qumsiyah (1985) examined all Egyptian specimens assigned to this species and identified them as Mehely's Horseshoe Bat. However, the Mediterranean Horseshoe Bat occurs in the region and the possibility of it turning up in Egypt cannot be excluded. Details of the noseleaf and sella are thus included.

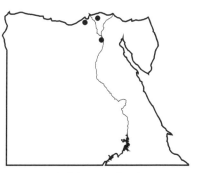

Mehely's Horseshoe Bat
(Rhinolophus mehelyi)

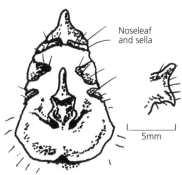

Noseleaf and sella

5mm

Mehely's Horseshoe Bat
Rhinolophus mehelyi

Similar species: Other horseshoe bats. See diagrams of sellae and noseleaves.

Leaf-nosed Bats—Family Hipposideridae

61 species worldwide, 1, possibly 2, in Egypt.

Small bats, similar (and closely related) to the true horseshoe bats. Like the true horseshoe bats they lack a tragus and have proportionately larger ears. They differ from the horseshoe bats in their facial structure. They lack the clearly defined horseshoe on the noseleaf, lack a sella, and, instead of a lancet, have a broad, flattened extension to the noseleaf.

Persian Leaf-nosed Bat
(Triaenops persicus)

Trident Horseshoe Bat
(Asellia tridens)

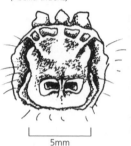

Noseleaves
of leaf-nosed bats

5mm

TRIDENT LEAF-NOSED BAT (TRIDENT HORSESHOE BAT, COMMON TRIDENT BAT) *Asellia tridens* (E. Geoffroy St-Hilaire, 1813)
Pl. 5

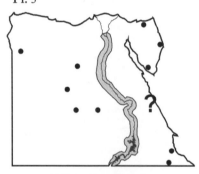

Trident Horseshoe Bat
(Asellia tridens)

Subspecies occurring in Egypt: probably *A. t. tridens*.
Arabic: *Khuffash azalya waraqi al-anf*
Identification: Length 66–89mm; Tail 18–28mm; Forearm 45–53mm. Male consistently larger than female. Differs from true horseshoe bats in structure of the noseleaf with 3 fleshy extensions on the upper surface rather than the lancet.
Ears relatively large and longer than in horseshoe bats. Tragus absent. Tail extends beyond flight membrane by up to 5mm. Color very variable but generally pale gray, paler below. Appears very pale when seen in flight at night.

Range and status: Much of Africa, though not the south. Arabia, Asia Minor east to Pakistan. In Egypt, a wide-ranging species recorded from North and South Sinai, Cairo and its environs, Saqqara, Fayoum, and south along the length of the Nile Valley to the Sudanese border. Also the Western Desert including Kharga, Dakhla,

Farafra, Bahariya, and Siwa, and along the Red Sea coast of the Eastern Desert.

Habitat: A species of desert and semi-desert. Roosts in caves, cliff crevices, ruins, old buildings, and temples (large roost at Dendera Temple).

Habits: Colonial species. Colonies large, hundreds (even thousands) of individuals. Flies late in the evening with rapid, twisting, turning flight likened to that of a butterfly. Flies low. Elsewhere, range thought to be migratory and numbers at roosts fluctuate. May also hibernate. Generally 1 young. Gestation 9–10 weeks. Reportedly preyed on by Sooty Falcon *Falco concolor*.

Notes: The Persian Leaf-nosed Bat *Triaenops persicus* has been reported in Egypt but the records were rejected by Qumsiyah (1985). This species is found over much of East Africa but no further north than the horn of Africa and in Arabia only occurs in the south and southeast. It is, therefore, unlikely to occur in Egypt. It differs from the Trident Horseshoe Bat in the structure of the noseleaf and in the tail not protruding beyond the interfemoral membrane. Emerges earlier in the evening than the previous species.

Similar species: True horseshoe bats. In the hand, the different noseleaf is diagnostic. Also note the tail protruding from the flight membrane. In flight appears paler. More widespread than any of the horseshoe bats.

Vesper Bats—Family Vespertilionidae

319 species worldwide, 10 species in Egypt.

A very large and complex group of bats; the number of species is constantly changing as new species are described. In Egypt, mostly small bats split into several distinct groups. Because of the size of the group, it is difficult to generalize about their features. However, they tend to be small bats with reduced eyes but distinct, well-separated ears with tragi that can be important for identification within genera. The heads are generally mousy and lack the complex noseleaves of other families. Pelage is generally short and dense. Tail either completely enclosed within interfemoral membrane or only extends a short distance beyond.

Right tragi of *Pipistrellus* spp.

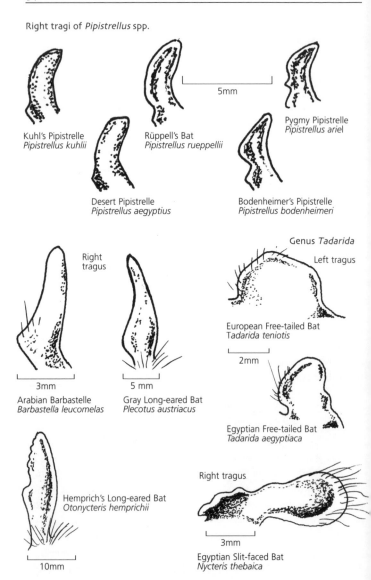

Kuhl's Pipistrelle
Pipistrellus kuhlii

Rüppell's Bat
Pipistrellus rueppellii

5mm

Pygmy Pipistrelle
Pipistrellus ariel

Desert Pipistrelle
Pipistrellus aegyptius

Bodenheimer's Pipistrelle
Pipistrellus bodenheimeri

Right tragus

Genus *Tadarida*

Left tragus

European Free-tailed Bat
Tadarida teniotis

2mm

3mm

Arabian Barbastelle
Barbastella leucomelas

5 mm

Gray Long-eared Bat
Plecotus austriacus

Egyptian Free-tailed Bat
Tadarida aegyptiaca

Hemprich's Long-eared Bat
Otonycteris hemprichii

Right tragus

3mm

Egyptian Slit-faced Bat
Nycteris thebaica

10mm

Tragi of various bat species.
With reference to Qumsiyah (1985), and Harrison and Bates (1991).

Pipistrelle Bats—*Genus Pipistrellus*

A large group of very small to small bats with relatively very small heads. Very difficult to tell apart, even in the hand. Five Egyptian species, only one of which is at all widespread. If in doubt, refer to the ranges of the various species, though these should be used with care. For example, Kuhl's Pipistrelle *Pipistrellus kuhlii*, the most common species, has not yet been recorded south of Luxor, but is found in northern Sudan and extends south to South Africa. Habitat may be a better guide. In the hand, color of body and flight membranes important as are the size and shape of the tragi.

KUHL'S PIPISTRELLE *Pipistrellus kuhlii* (Natterer, 1819)
Pl. 6

Subspecies occurring in Egypt:
P. k. marginatus.
Arabic: *Bibistril kuli*
Identification: Length 75–92mm; Tail 32–42mm; Forearm 31–36mm; Weight c. 5–8.5g. Small bat but the largest pipistrelle in Egypt. Wing membrane has a narrow, white margin running from the foot to the fifth digit, but this can be obscure in pale specimens where whole membrane is pale, and appears to dull with age. Also, less distinct in desert specimens. Tail moderately long but entirely surrounded by membrane, except for last vertebra which projects beyond. Ears broad, narrowing toward the tip. Tragus tall and narrow, slightly broader at base. Color variable. Egyptian specimens recorded from russet to olive-brown to pale brown, paler below. Gets paler with age.
Range and status: Much of Africa except Sahara and rainforest areas. Southern Europe, Middle East, and

Kuhl's Pipistrelle
(Pipistrellus kuhlii)

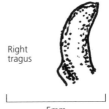

Right tragus

5mm

Kuhl's Pipistrelle
Pipistrellus kuhlii

Arabia east to Pakistan. In Egypt, very common especially in popu-
lated areas including cities. Recorded from Sallum and Mersa
Matruh on the north coast, and North Sinai. Found throughout the
Delta, inc. Lake Manzala, Cairo and its environs, inc. Saqqara,
Fayoum south along Nile Valley, with one questionable record from
Luxor. Not recorded south of Luxor.

Habitat: Roosts virtually anywhere including buildings on busy
streets. Also caves, farms, roofs, tombs, crevices, and cracks in walls.
Not a desert species.

Habits: A colonial species, colonies said to be detectable by
buzzing and squeaking on approach. Individual colonies small, up
to a dozen animals but may be more. Tends to emerge early and fly
moderately high, skimming along walls. Flight rather fast, direct,
and acrobatic. In Egypt, flies throughout year accumulating fat
reserves in autumn. Does not hibernate but may reduce activity.
Females give birth to normally 2 young in spring (May).

Notes: In some areas of its range, including Libya, Kuhl's Pipistrelle
is thought to be migratory. No evidence of migration has been found
in Egypt.

Similar species: Other pipistrelles. Kuhl's is the largest but in the field
this is not obvious. Range is a good clue, though see introduction to
group. Desert Pipistrelle *Pipistrellus aegyptius* is not recorded north of
Luxor and the Pygmy Pipistrelle *Pipistrellus ariel* has not been recorded north of Aswan. Bodenheimer's Pipistrelle *Pipistrellus bodenheimeri*
is only known from St. Katherine where Kuhl's is unrecorded. Of the
pipistrelles, only Rüppell's Bat *Pipistrellus rueppellii* overlaps and this is
a highly local species that is not associated with human habitation.

DESERT PIPISTRELLE *Pipistrellus aegyptius* Fischer, 1829.
Pl. 6
Subspecies occurring in Egypt: *P. a. aegyptius*.

Identification: Length 69–81mm; Tail 31–38mm; Forearm
29–33mm. Although there is some overlap in proportions with the
preceding species, this is a more slenderly built pipistrelle. Generally
lighter in color than Kuhl's Pipistrelle (which can, however, be very
variable), buffish brown above, lighter below. Probably only safely
distinguishable on skull characteristics.

Range and status: North Africa from Algeria, Libya, and Egypt. Also Upper Volta and possibly Kenya. In Egypt, only recorded from Luxor and Aswan. One record from Wadi Halfa in Sudan indicates that it may occur in Egypt south of Aswan.

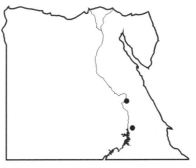

Desert Pipistrelle
(Pipistrellus aegyptius)

Habitat: Always associated with palm trees where they probably roost.

Habits: Very little known from Egypt. Probably gives birth in April/May.

Notes: In literature sometimes referred to as *P. deserti*, now treated as synonymous.

Similar species: Other pipistrelles for which range and habitat is important. See above for comparison with Kuhl's. Told from Pygmy Pipistrelle by larger size and dental characteristics.

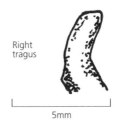

Right tragus

5mm

Desert Pipistrelle
Pipistrellus aegyptius

PYGMY PIPISTRELLE (EGYPTIAN DESERT PIPISTRELLE) *Pipistrellus ariel*
Thomas 1904
Pl. 6

Subspecies occurring in Egypt: unknown

Identification: Length 68mm; Tail 34mm; Forearm 31mm. A small pipistrelle, brown to deep olive-buff in color. Wing membranes pale brown. Ears narrow and pointed. Tragus distinct, curved inward with double notch

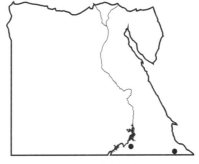

Pygmy Pipistrelle
(Pipistrellus ariel)

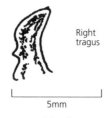

Right
tragus

5mm

Pygmy Pipistrelle
Pipistrellus ariel

at base. Probably indistinguishable from Desert Pipistrelle in field.

Range and status: A very rare (or rarely recorded) bat with scattered records from Sudan, Egypt, and Israel. In Egypt, very rare and known only from the Sudan Government Administration Area and Lake Nasser.

Habitat: Nothing known.

Habits: Nothing known.

Similar species: Other pipistrelles. In Egypt, only the Desert Pipistrelle occurs in the current species' known range. Probably only distinguishable on skull and dental characteristics.

BODENHEIMER'S PIPISTRELLE *Pipistrellus bodenheimeri* Harrison, 1960
Pl. 6
Monotypic

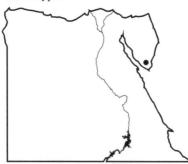

Bodenheimer's Pipistrelle
(Pipistrellus bodenheimeri)

Identification: Length 66–72mm; Tail 34–36mm; Forearm 30–31mm. A very small, delicate pipistrelle but with a relatively long tail, the last vertebra of which emerges from the flight membrane. Fur long, very pale in color, almost white, tinged with beige. Wing membrane uniformly dark while tail membrane pale and translucent. Ears pale and almost translucent, relatively large, broad, and rounded. Tragus distinct, almost triangular in shape, wide in the middle and extending half the height of the earlobe.

Range and status: Sinai, southern Israel, and Yemen. Described relatively recently and very little known.

Habitat: Little known but probably desert oases. Has been found in areas of cultivation surrounded by sandy desert, wadi floors, etc.

Habits: Little known. Flight low and delicate, generally around veg-

etation, e.g., tamarisk and eucalyptus. Gives birth probably in spring.
Similar species: Other pipistrelles. Bodenheimer's Pipistrelle is strikingly pale, except for wing membranes. In the hand, tragus shape distinct. No other pipistrelle recorded from east Sinai, though Kuhl's occurs in the north and the rare Pygmy Pipistrelle has been recorded just across the border in Israel.

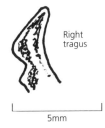

Right tragus

5mm

Bodenheimer's Pipistrelle
Pipistrellus bodenheimeri

RÜPPELL'S BAT (RÜPPELL'S PIPISTRELLE) *Pipistrellus rueppellii* (Fischer, 1829)
Pl. 6

Subspecies occurring in Egypt: unclear though Sinai specimens assigned to *P. r. rueppelli.*
Identification: Length 77–85mm; Tail 33–39mm; Forearm 31–34mm; Weight c. 7g. Medium-sized, rather broad-winged pipistrelle. Wing membranes distinctive in that they are pale gray contrasting with blackish limbs. Thumb 4.3mm with claw. Tail blackish. Muzzle and ears blackish, tragus relatively small reaching less than half the height of the earlobe. Fur fine, gray-brown above. Fur below pure white to the roots.
Range and status: Distribution throughout range discontinuous. Much of Africa south to Angola and Botswana. Also Arabia and Iraq. In Egypt, very few records from Nile Valley, inc. Cairo, Giza, Fayoum, and Qena.

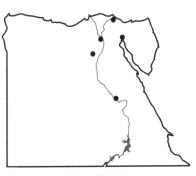

Rüppell's Bat
(Pipistrellus rueppellii)

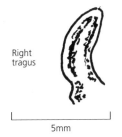

Right tragus

5mm

Rüppell's Bat
Pipistrellus rueppellii

Also recorded from Suez area. Three individuals recorded from Lake Manzala.

Habitat: Little known. Desert margins often in the vicinity of water, e.g., ponds or canals. Most specimens taken from around human habitation. In southern Africa, restricted to riverine forests. Not a desert bat.

Habits: Probably highly specialized. Most specimens in Egypt have been found while overturning boulders in desert and semi-desert areas looking for reptiles, etc. Appears to be solitary.

Similar species: Other pipistrelles. Pure white belly, including hair bases, contrasting with dark ears, muzzle, and wing membrane is diagnostic in the hand.

Serotine Bats—Genus Eptesicus
Externally very similar to the pipistrelles and only separated on skeletal characteristics, particularly features of the skull and penis bone. See Harrison and Bates (1991) for these details relating to the only Egyptian species.

BOTTA'S SEROTINE BAT *Eptesicus bottae* (Peters, 1869)
Pl. 7

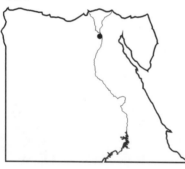

Botta's Serotrine Bat
(Eptesicus bottae)

Subspecies occurring in Egypt: *E. b. innesi.*

Identification: Length 92–97mm; Tail 39–43mm; Forearm 40–42mm. Medium-sized serotine. Ears moderately large and broad. Tragus relatively short and narrow. Tail moderately long and within membrane except for terminal 3–4mm. Fur is dense and moderately long. Buff above and white below only slightly tinged beige. Border at neck is obscure. All hair dark gray at base. Flight membranes reddish brown.

Range and status: Egypt, Israel, southern Arabia east to Iraq,

Afghanistan, and Pakistan. Egypt represents the western limit of this bat's range. It is very local and has only been recorded from Cairo and Giza. Recorded from South Sinai in 2006.

Habitat: Very variable. Has been recorded elsewhere from cultivated areas with eucalyptus and tamarisk in desert, gardens, date groves, ponds and riversides, and rocky mountains to 2,100m. Not recorded from true desert. Has been found roosting in buildings.

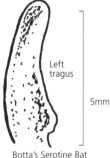

Left tragus

5mm

Botta's Serotine Bat
Eptesicus bottae

Habits: Little known in Egypt. Flight reported to be high and strong. Call audible. Flies relatively late after dusk. Breeding uncertain but females may give birth to up to 2 young in May.

Similar species: Pipistrelle bats are much smaller.

The following four genera are represented by only a single species each in Egypt, so group characteristics are discussed under each species description.

HEMPRICH'S LONG-EARED BAT *Otonycteris hemprichii* Peters, 1859
Pl. 7

Subspecies occurring in Egypt: *O. h. hemprichii* and *O. h. jin.*

Arabic: *Khuffash hambrish*

Identification: Length 112–122mm; Tail 51–56mm; Forearm 57–59mm; Wing-span 420mm. A large, stocky bat with huge ears (35–42mm). Ears huge, elongated though broad with moderately rounded tips. They do not meet on the forehead. Tragus large, elongated, and simple with serrated outer edge. Tail enclosed within membrane except for terminal 4–5mm. Calcar extends along half of tail membrane. Wings 'thick and leathery,' though appear pale in flight. Thumb large. A very pale bat, tinged buff above, pure white below with demarcation on neck reasonably distinct. Flight membranes virtually naked, ocher toward body paling toward wing tips, and translucent. Ears pale ocher. Subspecies *O. h. jin*, only from North Sinai, slightly larger than nominate subspecies but probably indistinguishable in the field.

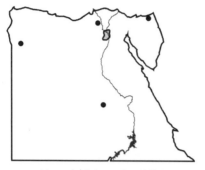

Hemprich's Long-eared Bat
(Otonycteris hemprichii)

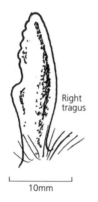

Right
tragus

10mm

Hemprich's Long-eared Bat
Otonycteris hemprichii

Range and status: North Africa from Morocco east to Egypt. Arabia, Turkmenistan, Afghanistan, and north-western India. In Egypt, *O. h. hemprichii* recorded from Western Desert, Siwa, Kharga, Wadi Natrun, Cairo and environs, margins of Fayoum, one record from Red Sea coast near Quseir. *O. h. jin* only recorded from North Sinai (al-Arish). According to Qumsiyah (1985) species probably occurs throughout Egypt.

Habitat: A desert species that can be found in very arid areas. Roosts in rock crevices but also found in old buildings.

Habits: Little known due to the preferred habitat of the species. Elsewhere, birth estimated to be in June. Although this is a desert species, it has been recorded flying over water. Bats emit a buzz likened by Harrison and Bates (1991) to that of a bumblebee. Remains found in Barn Owl pellets.

Notes: Much is still to be learnt about this bat in Egypt due to the fact its preferred habitat is little visited or researched as compared to the Nile Valley or Delta.

Similar species: Can only be mistaken for other large-eared bats. Gray Long-eared Bat is much smaller with the ears meeting on the forehead. It is not a desert species (see habitat description). Egyptian Slit-faced Bat has tail entirely enclosed by flight membrane and groove down face. It is largely confined to the Nile Delta and Valley (see habitat description). In *Tadarida* bats, the tail extends well beyond flight membrane.

SCHLIEFFEN'S BAT *Nycticeinops schlieffeni* (Peters, 1860)
Pl. 7

Subspecies occurring in Egypt:
N. s. schlieffeni.

Identification: Length 69–
79mm; Tail 28–33mm; Fore-
arm 29–33mm; Wingspan
180mm; Weight 4.5g. A tiny
bat, delicately built. Tail relatively short and almost entirely enclosed within flight
membrane. Ears fairly large,
broad with tragus about half
the length. Hair fine, longer

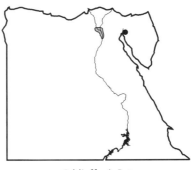

Schlieffen's Bat
(Nycticeinops schlieffeni)

(c. 4.5mm) above than below. Color very variable though paler below
and hairs always bicolored. Flight membrane uniform dark brown.

Range and status: Much of the savanna areas of Africa south to
Botswana and South Africa, and north to Sudan and Egypt. Also
southwestern Arabia. In Egypt, only recorded from Cairo and environs
(Saqqara) and Suez.

Habitat: Little known. Elsewhere, recorded from vicinity of gardens
and buildings. In southern Africa, in open woodland.

Habits: Little known. Said to fly low with a jerky flight. Hunts in
early evening. Not colonial, roosting alone. Little known of breeding
but twins have been recorded.

Notes: Due to its similarity to the pipistrelles and its wide distribution
in Africa, this species may prove to be more common in Egypt than
records suggest. Records of Savi's Pipistrelle *Pipistrellus savii* in Egypt
have all been assigned to Schlieffen's Bat (see Qumsiyah [1985]). Best
distinquished from smaller pipistrelles by dental characteristics.

Similar species: The pipistrelles, especially Bodenheimer's
Pipistrelle and Pygmy Pipistrelle neither of which overlaps in the
known range of the present species. Probably indistinguishable in
the field. In the hand, look at the tragus shape and the degree to
which the tail emerges from the flight membrane. Certain
identification from dental characteristics. Schlieffen's Bat has only
one pair of upper incisors and small, upper premolars absent.

ARABIAN BARBASTELLE (SINAI BARBASTELLE, EASTERN BARBASTELLE)
Barbastella leucomelas (Cretzschmar, 1826)
Pl. 7

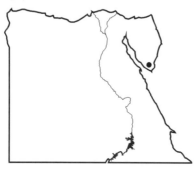

Arabian Barbastelle
(Barbastella leucomelas)

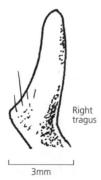

Right
tragus

3mm

Arabian Barbastelle
Barbastella leucomelas

Subspecies occurring in Egypt:
B. l. leucomelas.

Identification: Length 45–51mm; Tail 19–20mm; Forearm 38–39mm. Rather small, long-limbed bat. Ears forward-facing and rather large, though not elongated, joined at the forehead. Tragus large and hairy, elongated and pointed, more than half the length of the ear itself. Tail barely projects beyond flight membrane. Snout distinctive for densely haired swellings on either side, otherwise truncated and wide. Overall impression dark. Blackish above but shoulder hairs tipped buff. Below browner but with blackish throat paling toward the abdomen as pale tips to hair become more prominent. Hair longer above than below. Flight membrane uniformly pale brown. Thumb markedly small (4mm with claw).

Range and status: Eritrea, Sinai, southern Israel. Arabia east to Afghanistan, China, and Japan. In Egypt, only recorded from St. Katherine.

Habitat: Little known. Has been recorded from old buildings.

Habits: Little known but presumably similar to those of the Barbastelle *Barbastella barbastella,* which flies early, usually twice a night. Flight moderately fast and fluttering though sometimes with slower wing beats. Flies low. Barbastelle is generally solitary and no colony of Arabian Barbastelle has been found in Egypt.

Notes: Considered by Qumsiyah (1985) to be a subspecies of the Barbastelle *Barbastella barbastella leucomelas.*

Similar species: Range eliminates many species. Pipistrelles have narrower ears and lack facial swellings. Long-eared bats (*Otonycteris* and *Plecotus*) have much larger ears. Schlieffen's Bat has different ear shape, is significantly smaller, and has dark brown wing membranes.

GRAY LONG-EARED BAT *Plecotus austriacus* (Fischer, 1829)
Pl. 7

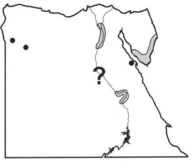

Subspecies occurring in Egypt: *P. a. christiei.*
Arabic: *Khuffash tawil al-udhun li-shamal Ifriqiya*
Identification: Length 92–96mm; Tail 42–48mm; Forearm 37–41mm. Small, pale, long-eared bat. Ears huge (only slightly shorter than forearm) and meet on the forehead. Tragus long and unlobed or notched, reaching half the length of the ear.

Gray Long-eared Bat
(Plecotus austriacus)

Unusual in that nostrils open upward. Tail relatively long with only very tip emerging from the flight membrane. Longhaired, especially above, and pale. Above, pale buff, though hair with dark bases. Below, hair white with dark bases. Demarcation at neck not clear.

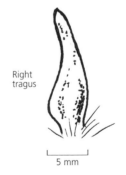

Right tragus

5 mm

Gray Long-eared Bat
Plecotus austriacus

Range and status: Much of Europe. North Africa south to Senegal and Ethiopia. Also Arabia east to Mongolia and western China. In Egypt, Nile Valley from Cairo and environs, inc. Saqqara, south to Aswan. Also Siwa in Western Desert, and Sinai, north to Nakhl.
Habitat: Seems to show a preference for roosting in dark areas such as pyramids, caverns, old buildings, and monuments. Like many Nile Valley bat species, recorded from Dendera Temple north of Luxor.

Habits: Little known as only considered a separate species from the Long-eared Bat *Plecotus auritus* relatively recently. Specimens from Dendera found hanging from the walls rather than the ceilings. Birth probably in May/June.

Notes: Only widely accepted as separate from the Long-eared Bat fairly recently, a distinction based largely on the structure of the penis bone, but now generally recognized.

Similar species: See Hemprich's Long-eared Bat for distinctions from other large-eared Egyptian bats.

Free-tailed Bats—Family Molossidae

c. 91 species worldwide, 2 species in Egypt.

Medium-to-large, robust, insectivorous bats, with large, forward-pointing ears and tails projecting well beyond the interfemoral membrane. These bats are also known as mastiff bats due to their large, wrinkled, often upturned snout. The ears are always large, with sparse hair and, in some species (though not the Egyptian ones), are fused at the base or along part of their length. The tragus is generally small and broad. Wings are long and narrow and the flight generally rapid. The flight membranes are mostly naked and leathery. Most distinctive is the tail. The only other bats with comparable tail projections are the rat-tailed bats where the interfemoral membrane is virtually absent and the tail is very long and slender. In free-tailed bats, the tail projects from the membrane half or more of its length and is more robust. At rest, the rather thick, short tail does not curl over the bat as in the much smaller rat-tailed bats but sticks up at an angle. Free-tailed bats are more mobile on the ground than most bat species and can scuttle effectively for cover, even on vertical surfaces, if disturbed. Diet consists of insects caught on the wing. Records are relatively sparse, perhaps because of their habit of flying high and fast.

EUROPEAN FREE-TAILED BAT *Tadarida teniotis* (Rafinesque, 1814)
Pl. 3
Subspecies occurring in Egypt: *T. t. rueppellii.*
Arabic: *Abu burneta al-kabir*
Identification: Length 122–139mm; Tail 40–52mm; Forearm 54–64mm; Wingspan to 450mm. Large bat, together with following

species second only in size to the Egyptian Fruit Bat, but readily distinguished from it. Heavily built, large-limbed bat with very long, slender wings. Snout long and slightly upturned with distinctly ridged upper lip, each vertical lip with a fringe of short hairs. Otherwise snout naked. Ears very large (length 18–33mm) and forward and downward pointing, curling in at the margins. Not fused at base though appear very close and may touch toward tips. Tragus small and square. Tail long and up to three-fourths of its length extends beyond the flight membrane. Fur long, especially on throat. Slate-gray above, paler below, coat spreading onto wing membranes.

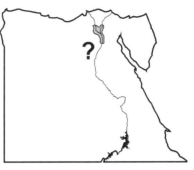

European Free-tailed Bat
(*Tadarida teniotis*)

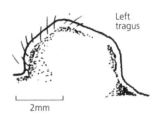

Left tragus

2mm

European Free-tailed Bat
Tadarida teniotis

Range and status: Southern Europe, North Africa, northern Arabia and eastward, as far as eastern China and Japan. In Egypt, only recorded from Cairo, Giza, Abu Rawash, and Saqqara, and probable record from the Fayoum.

Habitat: Roosts in old buildings and towers and found in Cairo itself. In Europe, largely known from urban areas. Also found in crevices inside caves, either deep and narrow, or taken from cavern roofs. Also in arid desert.

Habits: Little known. Flies fast and high. May be colonial. Estimated to give birth in mid-June but breeding also recorded later.

Similar species: For Egyptian Free-tailed Bat, see below. Distinguished from Egyptian Fruit Bat by large ears, free tail, smaller size, much smaller eyes not reflecting yellow-orange, and very different wing shape. Size and free tail separate it from other large-eared bats. Size and thick tail separate it from rat-tailed bats.

EGYPTIAN FREE-TAILED BAT *Tadarida aegyptiaca* (E. Geoffroy
St-Hilaire, 1818)
Pl. 3

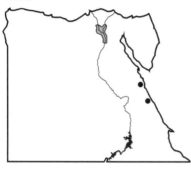

Egyptian Free Tailed Bat
(Tadarida aegyptiaca)

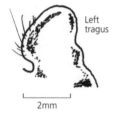

Egyptian Free-tailed Bat
Tadarida aegyptiaca

Subspecies occurring in Egypt:
T. a. aegyptiaca.
Arabic: *Abu burneta al-saghir*
Identification: Length 104–
120mm; Tail 41–46mm;
Forearm 47–56mm; Wing-
span 300mm; Weight 15g.
Large bat though distinctly
smaller than the European
Free-tailed Bat. Facial characteristics, including furrowed
snout and general form, similar to previous species though
tail proportionately generally shorter. Ears may be proportionately larger (18–23mm). Tragus similar: square
but with small, angular appendage. Fur fine and dense,
shorter below. Paler than European Free-tailed Bat,
grayish brown above, darker on back of head and back,
paler below, especially toward throat.

Range and status: Southern Africa, much of Central
and eastern Africa, Sudan, and Egypt, with isolated
population in northwestern Africa. East through
Arabia, Iran, Pakistan, India, and Sri Lanka. In Egypt, only recorded
from Giza area and Hurghada, and Qusair on the Red Sea coast.

Habitat: Semi-arid areas. Roosts in caves and large buildings.
Specimens from Abu Rawash obtained from crevices in caves.

Habits: Little known despite wide range but presumably as previous species. Elsewhere, lives in small colonies and gives birth to a
single young in summer. Flies high and rapidly. Active in winter.

Notes: Work on the Egyptian Free-tailed Bat has been limited; it
probably has a wider range than research indicates.

Similar species: See European Free-tailed Bat. Size and free tail
separate it from other large-eared bats. Size and thicker tail separate
it from rat-tailed bats.

The Carnivores — Order Carnivora

The carnivores are a large and diverse order of mammals, of which many, but by no means all, are adapted to a flesh-eating diet. The single characteristic that unites the order is the presence of four carnassial teeth. These teeth, which are designed to shear through flesh, are the last premolars in the upper jaw and the first molars in the lower jaw. Aside from this anatomical feature, the carnivores are a hugely varied assemblage of mammals. They range in size from the Weasel *Mustela nivalis* to the Polar Bear *Ursus maritimus*; they are found from the Arctic to the hottest deserts and the sea; and exploit food ranging from large herbivores, insects, fish, or (in the case of the famous Giant Panda) *Ailuropoda melanoeuca* bamboo.

In Egypt, the carnivores are amongst the hardest mammals to observe in the wild with the exception of the commensal Weasel and the opportunistic Red Fox *Vulpes vulpes*. Many have been drastically reduced in numbers due to hunting and habitat disturbance and, in the case of the Wild Cat *Felis silvestris*, possibly hybridization with domestic animals. Several species, such as the Cheetah *Acinonyx jubatus*, Leopard *Panthera pardus*, and Aardwolf *Proteles cristatus*, may already be extinct or very nearly so. The rarity of other species may sometimes be more apparent than real. The Sand Cat *Felis margarita*, for instance, inhabits areas of desert far removed from human activity and, thus, is very seldom encountered. Apart from rarity, another factor that makes the carnivores hard to observe is their nocturnal habits. Most species are active by night, especially in areas where there is human disturbance. The presence of carnivore species may only be apparent from droppings and prints.

Key identification features to look for in identifying any carnivore are general form, i.e., dog-like, cat-like, or musteline, as well as size, coat pattern, ear size and shape, snout shape, habitat, and location.

The Dogs and Foxes—Family Canidae

35 species worldwide with 6 species in Egypt.

The canids or dog family are the generalists of the carnivore order and include the wolves, jackals, and foxes. The canids are a relatively uniform group with long limbs, five toes on each forefoot (except the African Wild Dog *Lycaon pictus*), though the dewclaw is small and set high up, so track only shows four pads. The hind feet have four toes. The canids also have non-retractile claws, bushy tails, prominent muzzles, and large to very large, erect ears. Within this general pattern, they differ widely in size: in Egypt, they range from the Fennec Fox *Vulpes zerda* to the Wolf *Canis lupus*. A seventh species of dog, the African Wild Dog, was found in Egypt but became extinct in the early Dynastic Period. A late predynastic gray schist palette now in the Louvre in Paris depicts four African Wild Dogs along with two Giraffes *Giraffa camelopardalis*. This species is now probably Africa's most threatened canid and, in historical times, has been recorded no nearer than southern Sudan.

JACKAL (ASIATIC JACKAL, GOLDEN JACKAL, COMMON JACKAL) *Canis aureus* Linnaeus, 1758

Pl. 8

Subspecies occurring in Egypt: *C. a. lupaster*.

Arabic: *Ibn awi*

Identification: Length 101.2–127cm; Tail 29–34.7cm; Weight 10–15kg. A typically dog-like carnivore likened to a small, rather shaggy wolf. The head is like that of a domestic dog with a distinct and rather slender muzzle and relatively small, slightly pointed, erect ears. Legs long and slender. Tail relatively short and normally held below the line of the back. Coat is shaggy, individual hairs varicolored giving a grizzled or pepper-and-salt appearance, though looks grayish brown at a distance. There is a mane of longer hairs along the back. Ears are covered in much shorter hair and are rufous behind. Legs buffish with black stripe along the back of the foreleg. Inside of legs, throat, and belly whitish. Tail rather bushy, black along the top and at the tip. Does not touch the ground. One specimen from northwestern Sinai may relate to *C. a. syriacus*, which is smaller and more richly colored.

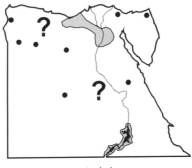

Jackal
(*Canis aureus*)

Range and status: North and East Africa, southeastern Europe through Turkey to the Middle East, southern Central Asia, Iran east through India as far as Thailand. In Egypt, from the Western Desert, including Siwa, Dakhla, and Kharga east to North Sinai, the Delta, Cairo and environs, including Gebel Asfar and Dahsur, Wadi Natrun, Fayoum, the Nile Valley south to Lake Nasser. Reported rare in the Fayoum, common along the shores of Lake Nasser including Wadi Allaqi. Widespread but probably declining in areas where it competes with feral dogs.

Habitat: Agricultural areas, wasteland and desert margins, rocky areas, and cliffs. Lakeshore at Lake Nasser. Not a desert animal except for the semi-arid northern coastal desert.

Habits: Largely nocturnal, especially where disturbed, but also reported active at dusk. Dens in natural caves, tombs, or dense scrub. Diet very varied. Omnivorous and opportunistic, recorded feeding on insects, snails, fish, chickens, young goats and sheep, as well as melons, watermelons, corn, small mammals, birds, and carrion. Has the habit of carrying off putrid or otherwise seemingly inedible items. Hearing and scent excellent. Sight good. Sociable, in packs or more often pairs. Mating occurs in early spring. Gestation c. 2 months, wild litters recorded in March, April, and May. 4–5 pups normal, up to 8 recorded. Very vocal. Voice a characteristic howl often followed by a short yelp delivered just after sunset and before dawn. Barks when excited, growls when annoyed, and female reported to utter a 'chak chak' with closed mouth as warning to pups.

Similar species: Wolf, see below. Feral domestic dogs (*baladi* dogs) tend to have shorter coats, be stockier, much more rufous with white patches, though extremely variable. The ears are proportionately larger and the tail often carried high, above the level of the back. Not normally black tipped. Fox spp. are smaller, slimmer with much larger ears and, in the two most common species, have a white-tipped tail.

WOLF (ARABIAN WOLF) *Canis lupus* Linnaeus, 1758
Pl. 8

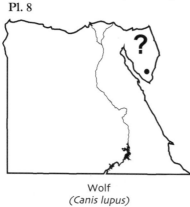

Wolf
(*Canis lupus*)

Probable subspecies in Egypt: either *C. l. pallipes* or *C. l. arabs*.
Identification: Length 114–140cm; Tail 31–45cm; Weight c. 25kg. Measure-ments are taken from Middle Eastern specimens. No Egyptian specimen has been measured. The following description is based on a single individual in emaciated condition found in al-Arish Zoo and reportedly captured in North Sinai. Typical large canine, much larger than previous species, long limbed with a rather narrow head. Coat generally short, longer on the back, neck, and tail but at no point shaggy. Fur grizzled gray, below more uniformly pale. Forelimbs lanky, tinged buff on upper leg to shoulder. Inside leg grayish. Hindquarters pale gray, tinged rust particularly at front of thigh. Head narrow, angular, and rather elongated. Forehead gray. Ears large and rather pointed, inside pale buff, tinged orange toward crown. Back of ears pale gray, tinged rust toward base. Upper side of muzzle gray, sides pale with black whiskers. Gray from muzzle continues around lower part of the eyes and up center of the head. There is a pale patch slightly above and inside each eye. Eyes roundish, amber-beige with dark brown pupils. Tail longer furred than most of body, grayish above, pale below with a black tip.

Range and status: North America, Europe (now patchy), Middle East including Arabia, northern and Central Asia to Japan. In Egypt, the Wolf is probably found in Sinai and there are more questionable, anecdotal reports from both the Eastern and Western Deserts. Classified as vulnerable by the IUCN and on Appendix I/II of CITES.

Habitat: Elsewhere in its range the Arabian Wolf has been recorded from desert margins. While the officials at al-Arish Zoo claim their specimen came from Wadi al-Arish, a relatively open, sandy area, local Bedouin claim Wolves were only found in the mountains of South Sinai where there is at least one recent sight record by reliable observers.

Habits: Not known in Egypt but probably solitary and nocturnal. Elsewhere the Arabian Wolf may be found in packs of up to 10 animals or in pairs or solitary. Permanent dens have not been recorded. Females give birth in March–May.

Notes: The existence of the Wolf in Sinai is open to much discussion (see Ferguson [1981]). The Egyptian Jackal *Canis aureus lupaster* is larger and longer limbed than the nominate subspecies of the Jackal, but smaller than the Arabian Wolf *Canis lupus arabs*. Ferguson examined the skulls of the Egyptian Jackal and considered it to be a small wolf. It is found in desert regions and should thus be smaller than the nominate subspecies that is found in more fertile areas; however, it is larger. It might thus be argued that it is more logical to consider *C. a. lupaster* as a small wolf rather than a large jackal. However, the specimen observed at al-Arish, and that photographed in Harrison and Bates (1991) are clearly different from the Egyptian Jackal. The current author considers them as separate—with the Arabian Wolf as a very rare resident of Sinai. Any further records of Wolves from Egypt should be as detailed as possible.

Similar species: Jackal smaller and shaggier, head more heavily furred, darker, and with comparatively smaller ears. Lacks the distinct head pattern of the Wolf with less white on the muzzle. Eyes closer set. Less long limbed. Domestic dog—see comparison with the Jackal.

RED FOX (NILE FOX) *Vulpes vulpes* (Linnaeus, 1758)
Pl. 9

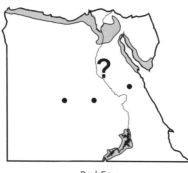

Red Fox
(Vulpes vulpes)

Subspecies occurring in Egypt: *V. v. aegyptica.*
Arabic: *Tha'lab ahmar*
Identification: Length 76.7–105.3cm; Tail 30.2–40.1cm; Weight 1.8–3.8kg. Largest Egyptian fox; female, on average, slightly smaller than male, and the most likely to be seen. The common name in Egypt is misleading, as the subspecies is not red. Ruddy to gray-brown above, darker on the back of the neck. Flanks grayer, tinged buff. Throat and belly dark even blackish, darker in winter, chin white. Forelegs brownish, marked with white and with black stripe down the rear side. Hind legs similar but with black limited to foot. Muzzle slender, beige above and reddish at the side with a dark streak running from muzzle to eye of varying distinctness. Whiskers black. Reddish brown below eyes, grayer on forehead. Ears large, inner side pale with long, whitish fringe hairs. Back of ear black. Tail full and bushy, paler below and with white tip. Juveniles paler and more uniform. Osborn and Helmy (1980) refer to two color phases, one brownish red and the other more yellow. Probably difficult to distinguish in the field. Desert specimens tend to be paler than those from the Delta and Valley. In early spring, foxes tend to look very shabby as they shed their winter coats.

Range and status: Much of Europe, North America, and Asia. From the Middle East across Iran, northern India to Japan. In Egypt, along the north coast from Sallum to Alexandria, the Delta, Wadi Natrun, the Fayoum, Cairo and its environs, including Saqqara, Abu Sir and Gebel Asfar, the Fayoum, Kharga, Dakhla, the northern Red Sea coast, Suez, and the Nile Valley south to the Sudanese border. Recently expanded into South Sinai and now resides in the Ras Muhammad National Park. A common species frequently encountered even at major sites such as the Step Pyramid. Expansion into South Sinai seems to be with the increasing spread of human activity. Reported to be rare in the Fayoum but sighted regularly. May go through periodic population declines, e.g., Saqqara 1995–96, possibly due to sacrophytic mange.

Habitat: Varied, though not a fox of true desert. Vegetated wadis, farmland, gardens, and desert margins.

Habits: Generally nocturnal but often seen during the day. In many areas, excavates an earth in the desert and comes down to the farmland to feed at dusk. Spends day in the earth or merely lying in a scrape in the shade but also sunbathes. Easy to approach on horseback. Earth may be burrowed in ground or the Fox may make use of tombs, ruined buildings, houses, etc. A wide range of food is taken, including insects, small rodents, fish, fruit, and vegetables. At Ras Muhammad, it digs for crabs. Must have daily access to water. Hearing, smell, and sight are all acute. Generally solitary but larger groups can be found in winter during the mating season, where groups of males may harass a female. Live as a pair during the rearing of the cubs. Litter size generally 3–5, born in February/March after c. 50-day gestation. Voice very variable. Most vocal during the rut when males have a triple bark. Also growls, chatters, and whines. Although much smaller than feral dogs (and presumably competing for the same food and habitat resources), foxes do coexist in some areas with packs of feral dogs. Dogs will chase foxes when they encounter them but the foxes are much faster and easily outrun the dogs. When cornered, will try to find a cavity inaccessible to the larger dogs.

Similar species: Distinguished from the Jackal by white-tipped tail, smaller size, and much larger ears. Rüppell's Sand Fox *Vulpes rueppelli* is much smaller and slimmer, with a pale belly and pale to rufous backs of the ears—never black. Also distinguished by habitat. Fennec much smaller and paler with huge ears. Blanford's Fox *Vulpes cana* with dark tip to the much bushier tail.

RÜPPELL'S SAND FOX *Vulpes rueppelli* (Schinz, 1825)
Pl. 9

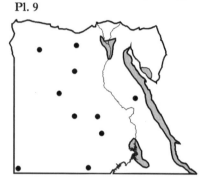

Rüppell's Sand Fox
(Vulpes rueppelli)

Subspecies occurring in Egypt: *V. r. ruepelli.*

Arabic: *Tha'lab rubil*

Identification: Length 68.4–90.6cm; Tail 27.3–38.7cm; Weight 1.1–2.13kg. Small, slim fox, much more lightly built than the previous species. Ears proportionately much larger. Legs shorter and slimmer than in Red Fox and pads almost completely concealed by hair (making tracks obscure). Fur is very fine and dense in contrast to coarse hair of the Red Fox. Reddish along the back, scattered with white-tipped hairs, buffish flanks and white beneath. Outside of limbs more rufous than flanks. Muzzle very slender, buffish above running up to yellow-buff forehead. Cheeks white. Diagnostic black patch in front of each eye and eyes surrounded by rufous ring. Whiskers black. Ears large, whitish within and buffish at back. Tail full and bushy, rufous-buff above, heavily flecked with black, paler at sides and below, and sometimes white-tipped.

Range and status: Desert regions of North Africa south to Somalia. Israel and Palestine, Arabia east to Iran, Afghanistan, and Pakistan. In Egypt, found throughout the Western and Eastern Deserts, and Sinai. Also around Lake Nasser in-cluding Wadi Allaqi. Largely absent from the Nile Valley but recorded from Wadi Natrun and the Fayoum. The most widespread fox in Egypt and the most likely to be seen in true desert. Reportedly rare in the Fayoum with sightings and possible tracks north of Lake Qarun in 1990.

Habitat: Much more of a desert animal than the previous species. Recorded from all types of desert, sandy and rocky, from semi-desert, and from rocky wadis. At oases, can be found in more vegetated areas, such as palm groves, and around wells. In the Fayoum, recorded from farmland and lake margins.

Habits: Largely nocturnal but also active by day and at dusk. May dig a shallow earth or just hole up in a rock crevice or dense vegetation. Omnivorous, feeding on small rodents, small birds, lizards, insects (such as beetles and grasshoppers), dates, and grass. Scavenged food probably only makes up a small portion of its diet. Reported by tribesmen in Gebel Elba to kill young lambs. Will drink when water is available, but can survive in areas where water is absent. Sight and hearing excellent, smell good. Adults generally found in pairs and may pair for life. Number of pups not known but probably 3 to 5. Variety of vocalizations including bark and angry yelp, also chattering. When antagonized said to arch back and raise tail spraying attacker with foul-smelling excretion from anal gland.

Notes: Research from Israel seems to indicate that Rüppell's Sand Fox is retreating in the face of agriculture and the subsequent arrival of the larger Red Fox. Certainly in Egypt there is little overlap between the two species, Rüppell's Sand Fox being largely absent from the Nile Valley and its margins. The Red Fox, however, requires water and cannot live in true desert.

Similar species: The Red Fox is larger with a dark belly, blackish in winter, and proportionately smaller ears with diagnostic black backs. The Fennec Fox is much smaller and paler with even larger ears and a black-tipped tail. Blanford's Fox lacks distinct facial markings and has a much fuller tail with a black tip. The Pale or Sand Fox *Vulpes pallida* has not yet been recorded from Egypt, but is found in northern Sudan and could conceivably turn up in the Gebel Elba region. It is similar to Rüppell's Sand Fox but has a black-tipped tail.

BLANFORD'S FOX *Vulpes cana* Blanford, 1877
Pl. 9

Subspecies occurring in Egypt: probably *V. c. cana*.

Identification: Length 73–76.2cm; Tail 32.4–36cm; Weight 710–956g. A small, strikingly beautiful fox with a particularly full and bushy tail. The coat is very thick and fine, pale gray on the flanks to pale russet below. Blackish from the base of the neck, along the back and down along the top of the tail. Legs are dark except for the front of the forelimbs, which are pale. There is a dark patch on each thigh. The muzzle is very pointed and the face narrow. As in Rüppell's Sand Fox, there is

Blanford's Fox
(*Vulpes cana*)

a dark patch between each eye and the upper lip, but this is narrower and clearer in Blanford's Fox. The forehead and crown are reddish brown. The ears are similar in proportion to those of Rüppell's Sand Fox and the backs are a similar shade of light brown. Feet small. Pads blackish and not hidden by hair as in Rüppell's Sand Fox. Tail strikingly full and bushy, blackish above, buffish russet on the side, flecked with black, and with a black tip.

Range and status: Iran, Iraq, Pakistan, and Afghanistan. Isolated populations in Arabia, Israel, and Sinai. In Egypt, only known since the early 1980s from southeastern Sinai and by very few specimens. It is not known whether the species is very rare or just extremely elusive. Listed on CITES Appendix II.

Habitat: Appears to be a mountain dweller, the naked pads being an adaptation to provide leverage on rocky slopes. In Egypt, only certain from the mountains of southeastern Sinai.

Habits: Little known but probably nocturnal though it has been observed during the day. Den is likely to be small clefts in the rock rather than a burrow. Uses rock shelves as rest sites. Feeds on insects and other invertebrates, small mammals, as well as vegetable matter, including grapes and melons. Litters of between 1 and 3 pups have been recorded born in April. Jumping ability impressive and will climb to escape, rather than run.

Notes: First specimen from Sinai was collected in 1977 and taken to Tel Aviv University where it was identified as Rüppell's Sand Fox. Subsequent examination, particularly of the naked soles of the feet, showed it to be Blanford's Fox. Subsequent specimens have been recorded from just across the border in Eilat.

Similar species: The Red Fox is much larger with black backs to the ears and white tip to the tail. Fennec much paler and not recorded from the same region. Rüppell's Sand Fox is similar in proportion but note black tip to extremely bushy tail in present species, more finely marked face, and darker limbs. There is a clear habitat difference. Blanford's Fox is a fox of cliffs and wadi slopes, Rüppell's Sand Fox of open wadi floors, and the Red Fox of more fertile areas.

FENNEC FOX *Vulpes zerda* (Zimmerman, 1780)
Pl. 9

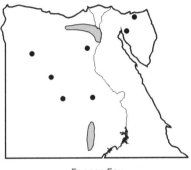

Fennec Fox
(Vulpes zerda)

Subspecies occurring in Egypt: probably *V. z. zerda.*
Arabic: *Fanak*
Identification: Length 52.3–61.7cm; Tail 18.6–23cm; Weight 1–1.5kg. Very small, uniformly pale fox with extraordinarily large ears. Very fine, soft fur, long and generally fairly uniform pale buff becoming paler still on the cheeks, chin, throat, and inside the ears. Variable but it is uncertain whether this is seasonal, sexual, or individual. Face darker buff but whitish around the eyes, which are large and dark. Flanks pale merging to white below. Muzzle pointed, very vulpine. Ears are huge, up to 10cm long, broad at the base, triangular, but rather rounded at the tip. Backs pale. Soles of feet covered with long hair obscuring the pads. Tail full and bushy, though not as full as in Blanford's Fox. Rufous above with very few black hairs. Tail tip black and more pointed than in other Egyptian foxes. There is a patch of black hair above the base of the tail marking a secretory gland.

Range and status: Much of North Africa south of the coastal desert from Morocco across to Sudan and Egypt, North Sinai, and Kuwait. In Egypt, recorded mainly from the Western Desert, inc. Wadi Natrun, Saqqara, Fayoum, Wadi al-Rayyan, Farafra, Dakhla, Kharga, and southeastern Western Desert. Isolated record from Sinai near Suez. Probably not common but several individuals have been seen for sale in Cairo pet shops. Reports of a pair breeding at Dendera

Temple probably refer to Rüppell's Sand Fox. Reported as rare in the Fayoum and only recorded from the desert there. Listed on CITES Appendix II.

Habitat: This is very much a desert fox, even more so than Rüppell's Sand Fox; it even avoids the more fertile coastal desert. Shows a preference for sandy desert with some vegetation.

Habits: Nocturnal, emerges at dusk and returns to its earth at sunrise where it spends the day. Earths in Egypt are recorded as a shallow and simple, single burrow—though deeper, more complex burrows have been recorded elsewhere. Omnivorous. Reported to feed on large insects, grubs, small mammals, birds, and lizards but with vegetation (such as berries and roots) making up a larger proportion of the diet than with most foxes. Does not need to drink but apparently will if water is available. As expected, hearing extremely good, sight and smell acute. Fennecs are sociable animals living in pairs or family units. Mate for life. Mating takes place in early spring. Gestation around 50 days and the female gives birth to up to 5 cubs in April to June. As with other foxes, has a wide range of vocalizations ranging from a high-pitched bark to a howling contact call.

Similar species: Other foxes but much smaller and paler. Distinguished from Rüppell's Sand Fox by much less black speckling especially in tail, by black tail tip, and gland at base of tail. Blanford's Fox lives in entirely different habitat, is much darker, and, in the hand, the soles of the feet are not obscured by hair.

The Weasel Family—Family Mustelidae
68 species worldwide with 4 species in Egypt

The weasel family, or mustelids, includes the weasels, polecats, martens, skunks, otters, badgers, and the Honey Badger *Mellivora capensis*. As can be seen above, it is not well represented in Egypt, though one additional species may prove to be found along the border with Israel, namely, the Honey Badger. The known Egyptian species are all slender, short-legged, agile carnivores. Three are strikingly patterned in black and white, or deep brown, white, and rufous, bearing a superficial resemblance to the new world skunks. The fourth, the Egyptian Weasel, is brown above, pale below, and restricted to urban areas.

One of the features of the weasels is their anal glands with which they mark their territories. The presence of weasels in a building, for instance, can often be detected by their strong odor, the scent being left with the droppings along their regular runways. A secondary function of the anal glands is in defense, where the discharge is used in much the same way as in the related skunks.

In identifying the Egyptian weasel species, key points to look for are size, color/pattern, and range.

EGYPTIAN WEASEL (WEASEL, LEAST WEASEL) *Mustela subpalmata* Hemprich and Ehrenberg, 1833
Pl. 10
Arabic: *'Irsa*
Identification: Larger and darker than European subspecies. Length (male) 36.1–43cm, (female) 32.6–36.9cm; Tail (male) 10.9–12.9cm, (female) 9.4–11cm; Weight (male) 60–130g, (female) 45–60g. A very small, slender carnivore. Males consistently larger than females. Long-bodied and short-legged. Head relatively small, snout broad, and ears small. Upper parts, legs, feet, and tail chestnut to dark brown. Underparts, including chin and throat, white to cream, which may or may not be clearly demarcated from the upper parts. Sometimes show brown

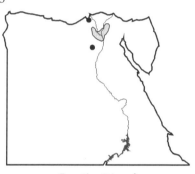

Egyptian Weasel
(*Mustela subpalmata*)

spots or blotches on the underside. Tail around one-quarter of total length, slender, not bushy, brown above and below, slightly darker at tip. In life, the overall impression is of a slender, elongated, brown mammal.

Range and status: Egypt, where restricted to the Nile Delta from Port Said to Alexandria south to Cairo, where it is very common. Also the Fayoum where reportedly common.

Habitat: In Egypt, largely a commensal of humans found in cities, towns, villages, and agricultural land.

Habits: Largely nocturnal but can be seen during the day. Most frequently encountered at night dashing across streets and disappearing beneath a parked car. Can be distinguished from rodents in these instances by the elongated body form, bounding gait, and relatively short tail. Nests in any cavity, pile of rocks, wall crevice, etc. Carnivorous, feeding on rats, mice, other small mammals, birds (up to the size of a pigeon), fish, and insects (including ants and cockroaches). Also scavenges waste from restaurants. Sight and hearing very good, but hunts mainly by scent. Solitary and territorial though male territory may contain female territory. Territory marked by urine and scat and smells strongly. Defends itself vigorously when cornered. Predators include dogs, cats, and larger birds of prey, even crows. Female gives birth to up to a dozen (normally 4–9) kits in lined den. Can have 3 litters in a year. Vocal with a variety of snorts, spits, yelps, and whines. Den with kits can be located by their piping.

Notes: Previously the Egyptian Weasel was considered to be a subspecies of the widespread Least Weasel *Mustela nivalis subpalmata*.

Similar species: The Striped Weasel *Poecilictis libyca* is larger, stockier, and distinctively patterned in black and white. The Egyptian Mongoose *Herpestes ichneumon* is much larger, grizzled gray, with a much thicker-based, longer tail.

STRIPED WEASEL (LIBYAN STRIPED WEASEL) *Poecilictis libyca* (Hemprich and Ehrenberg, 1833)
Pl. 10
Subspecies occurring in Egypt: *P. l. libyca*.
Arabic: *Abu mantan*

Identification: Length 40.5–47.2cm; Tail 16–19.3cm; Weight 500–750g. A small, long-haired weasel with distinctive coloring. Elongated body with full, bushy tail. Overall shaggy. Above and along flanks white, obscurely striped (stripes disjointed and irregular to the point that the animal can appear spotted) with black from behind the ears. Undercoat is black, long guard hairs pure white. Underside, legs, and feet are black. Head pattern distinctive. Muzzle black. White collar encircles the face running between the ears and eyes across the top of the head around to the underside of the mouth. Ears small and black, sometimes white tipped. Eyes dark. Tail bushy, patterned variably in black and white, but with underside of tip black.

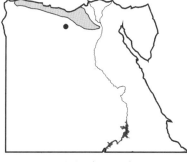

Striped Weasel
(Poecilictis libyca)

Range and status: A largely North African species found from Morocco east to Egypt, extending as far south as northern Nigeria. In Egypt, recorded largely from the western margin of the Delta and the coastal desert to Mersa Matruh. Also Wadi Natrun and the northeastern corner of the Qattara Depression. One isolated record of tracks from Libyan section of Gebel Uweinat. The species is found in Sudan so more southerly records might be expected.

Habitat: Sandy but vegetated desert and semi-desert. Vegetated sandy areas bordering the western Delta. Elsewhere, also recorded from rocky desert areas.

Habits: Largely nocturnal becoming active at dusk. Spends the day in den in rock crevice or burrow. Will use other animals' burrows or excavate one itself, a simple tunnel in sandy soil. Feeds on small mammals, birds, eggs, etc., though lizards have been reported to make up the bulk of its diet. Generally solitary and probably territorial. The Striped Weasel is not as agile as the Weasel. When threatened it will fluff up its hair to give the impression of greater size. If threat continues, said to present its rear with tail erect and spray the attacker with secretion from a stink gland. This behavior has been likened to that of the distantly related American skunks. Gives birth to 1–3 kits in spring, though earlier births recorded. Born naked or with fine, white hair, developing the adult coat pattern after about 3 weeks. Voice includes various spitting and growling when annoyed.

Similar species: The Zorilla *Ictonyx striatus* is the only other black- and white-striped mustelid confirmed in Egypt and only from the southeast. It is larger, with stripes much clearer and distinct. The band across the top of the head is broken. Marbled Polecat *Vormela peregusna* is deep brown, white, and rufous. From northeastern Sinai only.

ZORILLA (STRIPED POLECAT, ZORIL) *Ictonyx striatus* (Perry, 1810)
Pl. 10

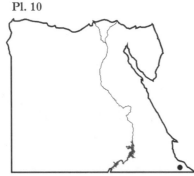

Zorilla
(Ictonyx striatus)

Probably monotypic
Identification: Length 61.3–64.2cm; Tail 27.3–28.8cm; Weight (male) 0.7–1.4kg, (female) 0.5–0.8kg. A strikingly marked black and white weasel, males generally larger than females. Elongated with rather bushy tail. Shaggy coated. Above white with three distinct black stripes running along the back joined at the head behind the ears and again at the base of the tail. Central stripe broadens toward the rear. After meeting at the base of the tail, the stripe continues along the top edge of the tail. Below, legs and feet black. Small, white ears with black tips. Band runs across forehead between eyes and ears, often incomplete. Claws of front feet long and strong. Tail shaggy, black at base and often at tip. Remainder black and white but proportions variable.

Range and status: Much of Africa south of the Sahara except for rain-forest areas. Extends north to Sudan and Egypt. In Egypt, recorded only from Sudan Government Administration Area (Wadi Darawena).

Habitat: Savanna, grassland, semi-desert, and scrub to 3,000m in mountainous areas. Avoids true desert and dense forest.

Habits: Largely nocturnal emerging at dusk. Spends day in burrow (self-dug or commandeered) or den in rock crevice or even build-ings. Food includes small mammals, birds, lizards, eggs, and large insects. Reported capable of killing large snakes. Generally solitary. Has stink glands and uses the same threat display as Striped Weasel. Gives birth to 2–3 young marked as the adults.

Notes: Care must be taken in oral reports since the Striped Weasel is also sometimes referred to as the Zorilla or Zoril (in French, the Striped Weasel is *le zorille* and the Zorilla, *le zorille commun*). Some authors recognize the subspecies *I. s. orythreae* from Egypt.

Similar species: Known records are separated geographically from the Striped Weasel but as both species occur in Sudan care should be taken with identification. The Zorilla is larger, much more distinctly striped, and with a differently marked tail. In the hand, it has 6 teats.

MARBLED POLECAT *Vormela peregusna* (Guldenstaedt, 1770)
Pl. 10

See notes on subspecies.
Identification: Length 59–65.5cm; Tail 16–17.8cm; Weight 265–520g. A very strikingly patterned mustelid. Elongated body with full, bushy tail. Facial region deep chocolate-brown with pale snout and a white band cir-cling the face above the eyes. Dorsum orange-buff and 'marbled' with dark brown clearly differentiated along

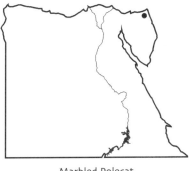

Marbled Polecat
(Vormela peregusna)

the flanks and uniform dark brown underparts. Considerable indi-vidual variation in the pattern of marbling. Legs proportionately short, also dark. Tail bushy, dark at base turning pale and dark brown at tip. Ears small and rounded, edged white.

Range and status: From Greece and Bulgaria in the west, through the Levant, and east through Iran, Afghanistan, and on to northern China and Mongolia. In Egypt, recently recorded from northeastern Sinai. Elsewhere a mammal of the steppes and decreasing as this habitat disappears and agriculture expands.

Habitat: Dry, open country including cultivated habitat, e.g., fields and orchards. Not a mammal of true desert.

Habits: Very shy. Nothing known of habits in Egypt but elsewhere largely nocturnal and crepuscular but sometimes active during the day. Den or burrow frequently is an enlarged rodent burrow. Food largely consists of small rodents, but also takes small birds, frogs, reptiles, and large insects. Generally solitary with a home range within which it is nomadic. Threat display dramatic, with tail fluffed up and held over the back. May also emit a foul odor caused by glandular secretions. Mating in spring, births following winter. Gestation short c. 23–45 days but implantation delayed. Litter size up to 8. Male plays no part in rearing the kits.

Notes: Recently recorded from Sinai and present across the border in Gaza. Harrison and Bates (1991) refer the Marbled Polecat from Arabia to *V. p. syriaca* to which the Egyptian specimens should probably be assigned.

Similar species: Can only be confused with the Striped Weasel and the Zorilla but both these species are all black and white with no brown or rufous. Ranges do not overlap. In the hand readily distinguished by 10 mammae (6 or 8 in the other two species).

The Civet and Genet Family—Family Viverridae
35 species worldwide with 1 species in Egypt.

The civets and genets, or viverrids, include not only the widespread civets and genets but also the palm civets, banded palm civets, and three single species subfamilies (limited to Madagascar), the Falanouc *Eupheres goudotii*, Fossa *Cryptoprocta ferox*, and Fanaloka *Fossa fossa*. They are a diverse group, presumably close to the ancestral line of carnivores, and range in form from the cat-like genets and linsangs, to the badger-like Binturong *Arctictis bin-*

turong. Often the coat is strongly patterned with spots and stripes or bands.

Superficially, the viverrids resemble the mustelids. They have short legs and long, rather slender bodies, rather pointed snouts, long tails, and small ears. Like the mustelids, they have highly developed scent glands. The single Egyptian viverrid, the Small-spotted Genet *Genetta genetta*, can be told from the superficially similar Egyptian Mongoose by its spotted coat, prominent ears, and very long, ringed tail. The coat and tail also differentiate it from Egypt's smaller cats.

SMALL-SPOTTED GENET (COMMON GENET, EUROPEAN GENET)
Genetta genetta (Linnaeus, 1758)
Pl. 11

Subspecies occurring in Egypt: *G. g. senegalensis*.
Arabic: *Ratam, Zariqa'*
Identification: Length 76– 104.4cm; Tail 35.2– 51.6cm; Weight 1–3kg. A very slen- der, cat-like carnivore. The body is very elongated, gray above, paler to buffish below. Patterned above with longitudinal stripes along the back and elongated spots along the flanks. Head with

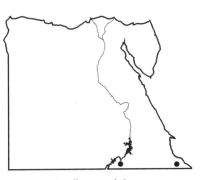

Small-spotted Genet
(Genetta genetta)

large, rather rounded ears, pinkish inside, back gray. Forehead gray with darker central stripe. Eyes brown. White patch below the eye contrasting with dark sides and base of muzzle. Muzzle whitish toward tip. Legs gray, paler inside. Tail very long and bushy, half total length, with up to 10 blackish rings on yellow-gray back- ground. Tip either dark or pale.

Range and status: South-eastern Europe (possibly introduced), northwestern Africa. Africa south of the Sahara north to Sudan and Egypt. Middle East and southwestern Arabia. In Egypt, very restricted distribution in southernmost portions of Eastern Desert with records from southeastern shore of Lake Nasser and from Gebel Elba region.

Small-spotted Genet
(*Genetta genetta*)

Habitat: Dry savanna, open acacia woodland to semi-desert. In mountains and in rocky areas with cover to 2,500m.

Habits: Little known in Egypt. Elsewhere, largely nocturnal but may emerge at dusk. Spends the day in burrow, tree, or rock crevice. Diet varied, largely carnivorous, recorded feeding on small mammals and birds, eggs, reptiles, frogs, fish, insects, spiders, scorpions, snails, etc. Also occasionally fruit and carrion. Stalks prey at night with body low to the ground and tail held horizontally out behind. Agile and can climb trees. Sight and hearing acute, smell good. Solitary or in pairs and probably territorial. When threatened, erects hair along back and tail and arches back to give the impression of a bigger animal. Can eject foul smelling secretion from anal glands. Gestation around 72 days, female giving birth to 1–4 kits in March to April. Voice recorded as 'uff-uff-uff' contact call, also hisses and spits when threatened. Kits whimper in nest.

Notes: The taxonomic status of the Small-spotted Genet is unclear. Harrison and Bates (1991) assign the Small-spotted Genet to *Genetta felina*.

Similar species: The only Egyptian small mammal that is spotted and striped in black with rings on the tail. Egyptian Mongoose is unpatterned with a proportionately shorter tail and separated geographically.

The Mongoose Family—Family Herpestidae
31 species worldwide with 1 in Egypt

The mongooses form a family closely related to the Viverridae and, indeed, are sometimes included in the same family. They are divided into two subfamilies; the Galidiinae with four species restricted to Madagascar, and the more familiar Herpestinae of Africa and Asia. They range in size from the Dwarf Mongoose *Helogale parvula* weighing in at just over 300g to the White-tailed Mongoose *Ichneumia albicauda* reaching 5kg. Most are grizzled gray, rufous, fawn, or yellow in color, rarely marked, and then striped, but never spotted as in the genets. The head is narrow and pointed and the ears generally very small. Most species are solitary, though some, such as the Dwarf Mongoose, the Banded Mongoose *Mungos mungo*, and the Suricate *Suricata suricata*, are sociable. Most species have a large anal gland in both males and females used in marking territory and in communication.

The Egyptian Mongoose can be distinguished from the mustelids by its larger size, grizzled coat lacking any black-and-white coloring, thick tail base (an impression given by the longer fur), and black legs.

EGYPTIAN MONGOOSE (LARGE GRAY MONGOOSE, ICHNEUMON MONGOOSE, ICHNEUMON) *Herpestes ichneumon* (Linnaeus, 1758)
Pl. 11

Subspecies occurring in Egypt: *H. i. ichneumen.*

Arabic: *Nims misri*

Identification: Length 90.9–106.8cm; Tail 36.3–46cm; Weight 1.9–4.0kg. Cat-sized, sharp-snouted carnivore like a large, shaggy weasel. Elongated body and short-legged. Above, uniform grizzled gray, grizzled appearance coming from the long guard hairs banded alternately dark

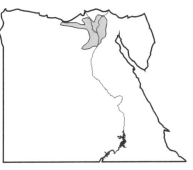

Egyptian Mongoose
(Herpestes ichneumon)

and pale. Below paler. Legs short, often barely visible beneath fur, colored as body with blackish feet. Muzzle sharp, blackish color. Rest of head as body, but hair shorter. Ears small, rounded and brownish. Tail long, grizzled gray, and with a long tuft of blackish hairs at the tip. Tail distinctively wedge-shaped with longest hairs at base getting shorter toward the tip.

Habits: Generally active by day but also at dusk and at night. Den may be a burrow (self-dug or adopted), rock crevice amongst rocks and boulders, or thickets of vegetation, such as reeds. Diet varied. Small mammals, birds, reptiles, including snakes, frogs and toads,

Egyptian Mongoose
(*Herpestes ichneumon*)

eggs, fish, crabs, large insects, fruit, and occasionally carrion. Famed for its ability to tackle poisonous snakes from which it is protected by the shaggy coat and a low sensitivity to poison. In ancient Egypt, recorded feeding on crocodile eggs. Sight, hearing, and scent good, though scent probably most important in finding prey. May rear up on hind legs to check the surroundings. Sociable, generally in pairs or family groups which probably hold territories. When threatened or excited, arches back and raises fur. Gestation 49–77 days. In Egypt, litters have been recorded virtually throughout the year, so probably no fixed breeding season, though in the Middle East most births recorded in spring. Litter size of 2–4 young. Rarely heard but has a wide vocabulary including a low whine, various chattering calls, and growling when annoyed.

Range and status: Southern Europe to Turkey and the Middle East. Throughout Africa except southwest and Sahara. In Egypt, restricted to the Nile Delta, including recent records from the shores of Lakes Burullus and Manzala, and Valley south to Cairo and its environs. Old records south to Asyut and beyond unconfirmed. There seems to be an expansion of the range across the northern coast to Mersa Matruh with the spread of agriculture. Reported to be common in the Fayoum.

Habitat: Agricultural areas with water. In the Fayoum, also recorded from settlements and from water margins, irrigation canals, etc.

Notes: Hafez (1993) includes the White-tailed Mongoose in his *List of Mammals of Egypt*. However, no basis for its inclusion is given nor any location. It has an extensive range throughout sub-Saharan Africa and southern Arabia and has been recorded in northern Sudan. Should it occur in Egypt, it would most likely be in the Gebel Elba region. It can readily be distinguished from the Egyptian Mongoose by the prominent white tip to the tail.

Similar species: For Small-spotted Genet, see previous species. The Egyptian Weasel is much smaller, more slender with short, brown hair. Tail is proportionately shorter and short haired toward base.

The Hyena Family—Family Hyaenidae

4 species worldwide with 2 in Egypt.

The hyenas are a small family of dog-like carnivores who are, however, probably most closely related to the viverrids. All four species are found in Africa with one, the Striped Hyena *Hyaena hyaena*, extending into Asia. Typically, they are thickset animals with far more powerful heads and forequarters, a sloping back, and weakly developed hindquarters giving them a peculiar, loping gait when running. The muzzle is short and broad, the jaws powerful and armed with large teeth for crushing bone (except the Aardwolf). The eyes and ears are large (as befits nocturnal predators) and the coat shaggy, often striped or spotted, with a mane along the back. In all but the Aardwolf, which retains five toes on the forefeet, there are four toes on each foot.

Although renowned for their scavenging habits, the three typical hyenas are also active predators and will also take eggs, fruit, insects, etc. The Aardwolf, while superficially very similar to the Striped Hyena, is much smaller and subsists almost entirely on termites, particularly harvester termites of the genus *Trinervitermes* and also termites of the genus *Hodotermes*. It is the distribution of these insects that likely controls the distribution of the Aardwolf. Its skull is far less massive than that of its relatives, and the teeth much weaker, reduced to peg-like structures with weak canines. Some authors have considered it sufficiently different to place it in a family of its own (the Protelidae).

The Striped Hyena has undoubtedly declined in Egypt. It is persecuted throughout its range and is the source of much local superstition. Furthermore, as the camel trains and caravans across the desert from Sudan are replaced by boat traffic and pick-up trucks, the camel carcasses that used to provide them with food have decreased.

In Egypt, the two hyaenids are most likely to be confused with dogs but are readily distinguished by their shaggy, striped coats, large, rounded ears, and long, bushy, black-tipped tails.

STRIPED HYAENA (BARBARY HYAENA) *Hyaena hyaena* (Linnaeus, 1758)
Pl. 12

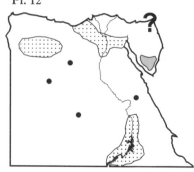

Striped Hyaena
(*Hyaena hyaena*)

Subspecies occurring in Egypt: *H. h. dubbah* (though subspecies in the Sinai not yet determined) and possibly *H. d. syriaca* or *H. h. sultana*.
Arabic: *Dab' mukhattat*
Identification: Length 131–142.5cm; Tail 29–35cm; Weight 19–45kg, though Egyptian specimens generally lighter. Large, rather front-heavy carnivore, superficially dog-like. Head proportionately large with broad, rounded ears, reverse blackish, paler inside. Muzzle broad, grayish, darkening toward tip. Throat dark. Forequarters powerful, hindquarters lower set and weaker. Hair long and shaggy with distinct mane from the neck to the base of the tail. Pale gray to beige with narrow brown to blackish vertical stripes, broken into spots on the neck and flanks. Legs pale with clear transverse stripes that break into spots on the inner side. Feet blackish. Tail relatively short, full, rather pointed at tip. Pale, darker above and with blackish tip.

Range and status: Much of North (where much reduced in numbers) and East Africa south to Tanzania. Sinai, the Middle East and Arabia, east to Iran, Pakistan, and south through much of the Indian Peninsula. In Egypt, historically widespread but everywhere much reduced or locally extinct. Formerly found along the margins of the Delta and Nile Valley along its entire length. Very rare in the Fayoum though fresh tracks reported in 1990. Also much of the Western Desert including all the major oases. Fewer records from the Eastern Desert but still found in the far south along Lake Nasser east along Wadi Allaqi. Mountains of South Sinai where a female and her cubs recently were photographed in the St. Katherine's Protectorate. Certain populations regarded as endangered by the IUCN.

Habitat: Desert margins along Nile Valley or around oases descending to cultivated areas at night. Formerly in desert areas along camel trains but has disappeared from these areas as caravans ceased. In Sinai, around settlements but heavily persecuted.

Habits: Nocturnal. Spends day in den in a natural cave, crevice, or overhang, which can be recognized by the accumulation of bones. Omnivorous, diet includes small mammals, birds, reptiles, fish, eggs, large insects, fruit (inc. dates, tomatoes, and watermelons), and carrion. Also human rubbish,

Striped Hyena den

animal bones, and even droppings. May hide excess food. Drinks when water is available but can reportedly survive without for extended periods; thus, in desert areas, food, not water, may be the limiting factor. Hearing and smell very good. Sight good. Solitary or in pairs. Not territorial but occupies a home range marked by scent. Reproduction in Egypt poorly known. Elsewhere, gestation 3 months giving birth in spring. Litter size 2–4 reared by both parents initially, female only after 6 months. Normally quiet but has a long, low howl when afraid or threatened. Also growls and whines.

Notes: As elsewhere, the Striped Hyena is subject to a great deal of suspicion, and hence persecution in Egypt. It has a reputation as a coward, as well as being savage, cunning, and gluttonous. And yet, it is killed not just for the inroads it reportedly makes on domestic livestock and crops, but for its body parts that reputedly give courage. Hyena parts are also considered variously as aphrodisiacs, medical treatments, and to counter the evil eye. Human persecution (by poisoning, shooting, and stoning), combined with the loss of the Western Desert camel trains, reduction in prey species, and habitat destruction, has greatly reduced this species in Egypt where it is now endangered. Protected areas in South Sinai and the southern Eastern Desert probably offer the Striped Hyena its last strongholds.

Similar species: The Jackal is much smaller and lighter, has a sharp snout, longer tail, and lacks the distinctive markings. The Aardwolf is similarly patterned, but is a much smaller, more delicate animal.

AARDWOLF *Proteles cristatus* (Sparrmann, 1783)
Pl. 12

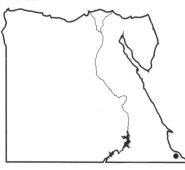

Aardwolf
(Proteles cristatus)

Subspecies occurring in Egypt: uncertain but that found in northeastern Sudan is *P. c. cristatus*.

Arabic: *Dib, 'Usbar*

Identification: Length 85–110cm; Tail 20–30cm; Weight 7–10kg. Like a delicate version of the Striped Hyena with a proportionately smaller head. Jackal-sized with large, rather pointed ears, and a slender muzzle. Muzzle black, as is chin and area around the eyes. Coat shaggy with erectile mane from the neck down the sloping back to the base of the tail. Color beige to grayish with a series of diagonally sloping stripes along the flanks. Legs striped as in the Striped Hyena. Feet blackish. Tail full and bushy with varying amount of black at the tip. Teeth are much reduced, molars and premolars peg-like, and canines slender. Very different from the powerful teeth of the Striped Hyena.

Range and status: Two main populations, one in East and central Africa, the other in southern Africa. In Egypt, only known from two specimens taken in the Sudan Government Administration Area. This is the northernmost limit of the Aardwolf's range and the population, if still present, is probably very small and may not even be permanent. Presumably, distribution depends on its very specific diet.

Habitat: Open dry plains and savanna. May be found in rocky areas, but avoids mountainous country.

Habits: Nothing known in Egypt and little studied elsewhere. Active at dusk and at night. Spends the day in a burrow that is generally adopted. Diet is highly specific feeding almost exclusively on termites (genus *Trinervitermes*) with small mammals, eggs, other insects, and carrion making up a very minor proportion of the diet. Sight and scent good, but hearing, with which it locates its prey, excellent. Solitary when feeding, but probably territorial in pairs. Territories marked by strong-smelling anal gland. When threatened, erects mane and exposes anal glands. Breeding very little known. Gestation 59–61 days with usually 2–4 young. More than one breeding female may occupy a single burrow. Generally silent, but may growl under threat. **Notes:** As noted in the family introduction, the Aardwolf is sometimes assigned to the family Protelidae of which it is the sole representative. **Similar species:** See Striped Hyena. The Jackal has no stripes, smaller ears, and its back does not slope down to hindquarters.

The Cat Family—Family Felidae

35 species with 6 species in Egypt. A seventh, the Lion *Panthera leo*, probably became extinct in late pharaonic times.

The cats, or felids, are the ultimate predators and the most strictly carnivorous of the carnivores, with the possible exception of the African Wild Dog. As a group, they are relatively uniform with all members of the order, from the diminutive Sand Cat to the powerful Tiger *Panthera tigris*, being instantly recognizable as cats. Typically they have large, mobile ears and large, forward-facing eyes. They hunt mainly by sight and sound, though scent is also well-developed. The muzzle is short and surrounded by long and very sensitive whiskers known as vibrissae. The teeth are powerful with long, sharp canines used in grabbing and securing prey. The tongue of the cats is also distinctive; it is covered with sharp papillae giving it the texture of rough sandpaper.

The limbs may be long and slender, as in the Cheetah, or short and sturdy, as in the Leopard. There are five toes on the forefeet, though only four appear in tracks, and four on the hind feet. The claws are sharp and often wrongly called 'retractile.' In actual fact, they are erectile, that is to say that they can be drawn out of their protective sheaths when needed, e.g., during a kill, rather than being withdrawn when not needed. Only in the Cheetah can the claws not be fully withdrawn. In most cats, the tail is long and important for balance, though in two Egyptian species, the Swamp Cat *Felis chaus* and the Caracal *Felis caracal*, the tail is relatively short.

The wild cats found in Egypt exhibit a wide range of hunting techniques. The Leopard ambushes its prey or approaches within a short distance after a stalk. The Caracal is renowned as a bird catcher, pouncing into a flock of birds and catching by surprise. The Sand Cat is a patient stalker while the Wild Cat and the Swamp Cat are more opportunistic. The Cheetah uses tremendous bursts of speed— up to 100kmph—to outsprint its prey in large, open areas which provide little cover for stalking.

Egypt's six wild cats are sufficiently different to present little problem in field identification though the Wild, Swamp, and Sand Cats are fairly similar. The main challenge lies in separating the true Wild Cat from tabby-type feral cats. The main features to look for are the color of the back of the ear, the shape of the tail, and the leg length, though none of these are easy to see given a quick glimpse. The situation is further complicated by hybridization around settled areas. The Leopard and the Cheetah are both extremely rare and the Sand Cat, Wild Cat, and Caracal very elusive. Probably the best chance of glimpsing a wild felid in modern Egypt is the Swamp Cat, which seems to have adapted better than most to human habitat disturbance.

WILD CAT *Felis silvestris* Schreber, 1777
Pl. 13

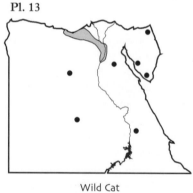

Wild Cat
(Felis silvestris)

Subspecies occurring in Egypt: *F. s. libyca* and *F. s. tristrami*.
Arabic: *Qitt gabali libi*
Identification: Length 61–93.5cm; Tail 23.7–39cm; Weight 2.5–6kg. Male on average larger and heavier than female. Like a large domestic cat and can be difficult to distinguish. Typical cat form but with longer limbs and with proportionately longer tail. Grizzled buffish above with blackish stripe down the center of the back. Paler along flanks. Whitish to pale buff below marked with pale brown spots that

form vertical stripes along the shoulder and sides. Brown transverse bands on the legs, legs paler inside. Feet yellowish with broken stripes. Muzzle short, orange around nose with white patch below the eye. Cheeks striped. Throat and upper lips also white. Discreet striping on face. Almost no tuft on ears, orange-buff behind, whitish inside. Tail long, blackish above, at base with three blackish rings and dark tip. *F. s. libyca* may be distinguished from *F. s. tristrami* (from which it is separated geographically) by smaller size and paler coloring  with less defined patterning and paler orange-buff behind ear.

Range and status: Central and southern Europe, most of Africa (except the Sahara and rainforest regions), Sinai and Middle East, east to Central Asia, Iran, Afghanistan, Pakistan, and India. In Egypt, *F. s. libyca* is found along the margins of the Delta and Nile Valley, including the Fayoum where reportedly common, and in the Mediterranean coastal desert to Mersa Matruh. Also recorded from the Western Desert at Bahariya and Dakhla and tracks recorded in northern Eastern Desert. *F. s. tristrami* is known from North Sinai around al-Arish and southwestern Sinai.

Habitat: Dry areas with cover including cultivated land in the coastal desert zone and elsewhere. Also rocky areas in mountains. Around settlements in the Fayoum, although the purity of the population might be questioned.

Habits: Little known in Egypt but probably strictly nocturnal. Lies up in tree, rock crevice, etc. Diet includes small mammals (up to the size of the hare), birds, reptiles, and insects. Hunts by hearing and sight, both of which are excellent. Scent less well developed. Solitary, male territorial range depending upon food supply. When threatened reacts as domestic cat, arching back and erecting fur, spitting and hissing, with claws extended. Agile, runs well and is a good climber. Nothing known of reproduction in Egypt. Elsewhere, gestation 56–60 days giving birth in spring to 2–5 kittens. Voice as in domestic cat.

Notes: The status of the Wild Cat in Egypt is difficult to ascertain because of confusion with the domestic or feral cat. Feral cats abound in many towns and villages, but the degree to which they interact with populations of Wild Cats in Egypt is not known. There is no confirmed record of a wild/feral hybrid specimen from Egypt. Some authors claim that the two interbreed, a claim also made by some Bedouin. In Sub-saharan Africa, hybridization with the feral cat is considered to present the biggest threat to the species and it is increasingly unlikely that pure populations exist around areas of human activity.

Similar species: The domestic cat may be very similarly patterned, but is smaller, has shorter legs, a shorter tail, and is generally less rufous behind the ears. Note also that the true Wild Cat is a solitary animal except when pairing. The Swamp Cat is larger with a proportionately much shorter tail, has ear tufts, and is virtually unmarked. The Sand Cat is smaller, paler, has huge ears, and paw pads completely covered with hair.

SWAMP CAT (JUNGLE CAT) *Felis chaus* Guldenstaedt, 1776
Pl. 13

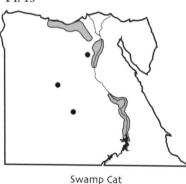

Swamp Cat
(Felis chaus)

Subspecies occurring in Egypt: *F. c. nilotica.*
Arabic: *Qitt barri nili*
Identification: Length 80.5–104cm; Tail 21–28cm; Weight 7–11.2kg. Form is that of a very large, short-tailed domestic cat. Above, grizzled yellow-brown paling toward flanks and with a darker stripe down the center of the back. Underside and inside of legs paler. Weakly patterned with pale spots on flanks and belly and pale stripes on hind legs. Feet ocher to brown. Much less strongly marked than previous species. Head with rather elongated muzzle, buffish along nose with white patch below eye and very faint markings along forehead. Cheeks unpatterned. Whiskers mixed black and white. Ears large with short, black tuft, russet with blackish base behind. Tail distinctly short with rounded tip, buffish with two blackish rings and black tip. Kittens striking-

ly short tailed with blue-gray eyes. Claws significantly more robust than those of domestic kittens.

Range and status: In Africa, only in Egypt with records from central Sahara now discredited. Also Middle East including Turkey to the Caucasus, east through Iran and Pakistan, to India and Sri Lanka, and on to Nepal, China, Burma, Thailand, and Vietnam. In Egypt, in the Delta, including recent records from Lake Manzala and near Damietta, and along the Mediterranean coastal desert to Mersa Matruh. South along the Nile Valley to Aswan with three recent records of kittens from cane fields in Luxor. Also recorded from Dakhla and Kharga in the Western Desert. Reported as common in the Fayoum. This species is still persecuted in Egypt, with trapping recorded along the Delta coast and pelts regularly for sale in the market at Kerdassa and the Khan al-Khalili. At the Khan, Swamp Cat pelts are sometimes painted with black spots and sold as African. It is protected by Egyptian law and is covered under CITES Appendix II.

Habitat: Areas with thick cover such as agricultural farmlands (cane, maize fields, etc.), marshes, and reed beds. In the coastal desert, found in areas of less deep cover and even recorded from sea cliffs along the north coast. Recorded from around settlements in the Fayoum. In Israel, recorded around fish ponds and tamarisk thickets. Probably the cat species that has adapted best to the disturbed habitat created by human activities.

Habits: Largely nocturnal but also active at dusk and occasionally during the day (presumably where not disturbed). Den in burrow (not normally self-dug), rock cavity, deep thicket, or dry area in reed bed. Also in abandoned buildings. Diet includes small mammals, birds, reptiles (including snakes), fish, and eggs. Has been known to attack livestock and poultry. Sight and hearing acute, scent good. Probably largely solitary occupying a home range, though in Israel, up to 3 adults have been seen together and females observed with up to two kittens. Males also reported helping in rearing young. When threatened, reacts in the same way as the Wild Cat, arching the back, raising the fur, and hissing. Reported to defend itself fiercely. Other calls include a loud mewing at mating and the familiar purr. Gestation 66 days. In Egypt, births recorded in January to April with 2–6 (generally 2–3) kittens reared only by the female or with help from male.

Notes: Although mummified by the ancient Egyptians, there is no evidence that the Swamp Cat was domesticated by them. The theory that Swamp Cats contributed to the present domestic cat is now generally rejected.

Similar species: Wild Cat, see previous species. The Sand Cat is much smaller, paler, with proportionately longer tail, and very different habitat requirements. The Caracal has a uniform coat, not grizzled, with black backs to the ears, and very long, prominent ear tufts.

SAND CAT *Felis margarita* Loche, 1858
Pl. 13

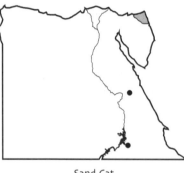

Sand Cat
(Felis margarita)

Subspecies occurring in Egypt: *F. m. margarita* though Sinai specimens have been assigned to *F. m. harrisoni.*

Identification: Length 66–83cm; Tail 23–31cm; Weight 1.5–3.4kg. Male generally larger and heavier than female. A diminutive cat, smaller and more compact than the Wild Cat with proportionately shorter legs. Pale buff above, paling toward flanks and legs, and white below. Throat tinged orange. Obscurely marked with pale brown stripes along the flanks. Forelegs strongly marked with two transverse blackish stripes. Hind legs less strongly marked with up to 5 brownish stripes. Feet with pads completely covered with long, brown hair. Head broad and flat, face pale marked with orange-buff. Ears very large and broad, almost meeting over the forehead. Inner side whitish, outer buff with darker patch. Eyes large with yellow iris. Tail long (about half the body length) grayish buff above, paler below. Faintly ringed with 2–6 tail rings and with clear blackish tip. *F. m. harrisoni* differs from nominate subspecies in 5–7 tail rings, pure white paws, and larger, broader skull.

Range and status: Deserts of North Africa, Sinai, Israel, and Arabia east to Iran and southern Central Asia. Everywhere rare and certain populations classified as endangered by IUCN. In Egypt, very little known but recorded from a few localities in the Eastern Desert. In the 1980s, there were records from the Qift–Quseir Road and Ras Abu Fatma in the Sudan Government Administration Area. Evidence points to there being a population of this elusive little cat in North Sinai. In 1995, an individual reportedly trapped in North Sinai was sent to al-Arish Zoo. Subsequent specimens obtained from northeastern Sinai in rocky terrain. Stuffed specimens occasionally turn up at Kerdasa—at least 4 since 1990—and one wildlife trader was known to be in illegal possession of 7 live Sand Cats in 1996, so it may be more widespread than the very sparse verified records suggest. Protected by Egyptian Law and under CITES Appendix II. Certain populations classified as endangered by IUCN.

Habitat: Sandy deserts for which the heavily haired soles are thought to be an adaptation. Also more rocky areas, but not mountains.

Habits: Little known. Strictly nocturnal, wanders widely at night though has been observed basking by day on very rare occasions. Has been observed surveying surroundings for up to 15 minutes with just its head out of the burrow prior to coming out. Similar procedure practiced before returning into the burrow. By day in a burrow or other shelter amongst rocks, etc. Dens may be used by several cats at different times. Not a good climber. Thought to feed largely on small desert rodents, including the Lesser Egyptian Jerboa *Jaculus jaculus*, Cairo Spiny Mouse *Acomys cahirinus*, and probably *Meriones* spp. Sand Cats observed on the Qift–Quseir Road were thought to be preying on *Meriones crassus* that had been attracted to the road by leaking grain from lorries. Also seen to take geckos (*Stenodactylus* spp.), but probably also takes birds, other reptiles, and large insects. Hunts largely by hearing, which is excellent. Sight very good, scent less well-developed. Solitary with home range probably large (owing to sparse food resources), though territories of males may overlap. Radio tracking in Israel has shown that Sand Cats will move up to 8km in one night, with an average of 5.4km covered. When threatened behaves much as

Wild Cat. Breeding unknown in Egypt but elsewhere gestation 63 days giving birth in March/April to 2–5 kittens. Voice as for Wild Cat but has a loud mewing mating call likened by one author to the yelp of a small dog. Tracks almost invisible because of dense fur. Pads not apparent.

Notes: In Israel, where the Sand Cat is also rare, the biggest threat to its future seems to be the spread of agriculture and the destruction of dune habitats allowing invasion by stronger, more aggressive species such as the Wild Cat and the Swamp Cat.

Similar species: See Wild Cat and Swamp Cat.

CARACAL *Felis caracal* (Schreber, 1776)
Pl. 14

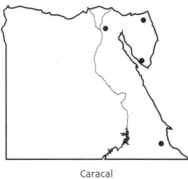

Caracal
(Felis caracal)

Subspecies occurring in Egypt: probably *F. c. schmitzi*
Arabic: *Umm rishat*
Identification: Length 85–120cm; Tail 20–30cm; Weight 8–18kg. Male generally bigger and heavier than female. A medium-sized cat characterized by its short tail and long ear tufts. Uniform orange-brown above, paler to white below with pale spotting. Striping on legs at best faint. Overall appears very uniform. Hindquarters set slightly higher than forequarters. Head rather angular and strikingly marked. Pale above each eye with a clear dark line running from above the eye down along the tear line and along to the nose. Chin and throat white. Ears large, long, and pointed with elegant, elongated black tufts up to 60mm long. Back of ears blackish and very important in social contact, animals using a form of 'semaphore.' Tail short and not black tipped.

Range and status: Most of Africa except the Sahara and rainforest areas. Sinai, western and southern Arabia, and Middle East, east to Iran, Afghanistan, Central Asia to India. In Egypt, little known

and rare. Recorded only from the
Eastern Desert and North and southwestern Sinai. Unconfirmed reports
of tracks in central Sinai. Highly elusive and may be more widespread
than the very few records indicate.

Habitat: Open country from sandy
desert to savanna, and open plains.
Also mountain areas.

Habits: In Egypt unknown. Elsewhere a very elusive cat, largely nocturnal but may be active at dusk or
during the day. During the day, lies
up in rock crevices, caves, burrows,
or dense bush. Renowned as a bird

catcher, grabbing a flushed bird from the air with a high leap. Agile.
Also feeds on mammals up to the size of a full-grown male gazelle
(killed in the same manner as the big cats by suffocation), reptiles,
and occasionally fresh carrion, including dead fish, and even fruit.
In the Negev, Cape Hares *Lepus capensis* and partridges (*Alectoris*
spp.) most commonly taken. Senses of sight and hearing acute.
Scent good. Generally solitary though occasionally in pairs or family groups. In Israel, up to 8 individuals were recorded around one
fishpond. Male probably territorial, size of range depending on
prey density. Gestation 69–70 days. In southern Arabia, birth is
estimated to be in early August. Nursery den lined with fur and
feathers. Litter size 1–6 (generally 2–3). Not vocal, but spits and
growls if threatened. Contact call is a sharp bark.

Notes: There are occasional reports in the Egyptian press of 'lions'
killed in Middle and Upper Egypt. The lion became extinct as a wild
animal in Egypt in ancient times. It is probable that such reports
relate to Caracals (Leopards are unrecorded in these areas and are
referred to as *nimr* not *asad*).

Similar species: No other Egyptian cat is so uniformly colored. The
Swamp Cat is smaller, has a ringed and black-tipped tail, much
smaller ear tufts, and a less strikingly marked face. The Leopard and
Cheetah both clearly spotted.

CHEETAH (HUNTING LEOPARD) *Acinonyx jubatus* (Schreber, 1776)
Pl. 14

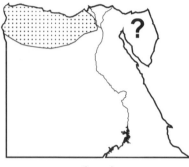

Cheetah
(Acinonyx jubatus)

Monotypic
Arabic: *Fahd siyad, Shita*
Identification: Length 175–220cm; Tail 65–80cm; Weight 40–60kg. A large, slender, small-headed cat, leopard-sized, but built on greyhound lines. Above, flanks and outside of legs buffish to yellow covered with solid, round, dark spots never arranged in rosettes as in the Leopard. Below, paler with more diffuse spotting. Fur generally short and dense, but longer below and with an erectile mane along the shoulders and back. In the cubs, this is larger and distinctly pale, a pattern said to provide protection through mimicry of the Honey Badger. Head proportionately small with flat top and small ears. Head densely spotted, chin white. Most distinctive feature is the blackish 'tear mark' running from the eye down the side of the face. Muzzle short. Tail long and full, broader at tip than base. Spotted above for the basal half, latter half ringed with up to 6 blackish rings, the final one being the broadest. Tail tip white. Claws only semi-retractile.

Range and status: Formerly over much of Africa outside the rainforests, now almost entirely south of the Sahara. Also formerly over much of the Middle East, including Arabia to Iran, Pakistan, and India. Now extinct in this region, except for a small population south of the Caspian Sea. Almost certainly extinct in Arabia. In Egypt, very rare, possibly extinct. The Egyptian population is probably the last remnants of the Cheetah north of the Sahara (though 2 were reportedly shot in Libya in 1980). Restricted to the Western Desert in and around the Qattara Depression. Its current population is unknown but very low. Hunting, habitat disturbance, and reduction in prey populations (e.g., Dorcas Gazelle *Gazella dorcas*) have brought the population down to a critical level. Records from North Sinai lack verification. Recent records include a female and 3 cubs shot by Bedouin hunters in 1993 and a female and 2 cubs in November 1994. No Qattara Cheetah has been photographed live, evidence coming from tracks and skins. Its subspecific status is unknown, but the Egyptian population may be genetically important for the long-term survival of the species (due to the genetic uniformity of the Cheetah as a species). Protected by Egyptian Law, listed by the IUCN as vulnerable, and on CITES Appendix I.

Habitat: Open plains and savanna to semi-desert. In Egypt, known from acacia groves in the Qattara Depression, the Mediterranean coastal desert, and, historically, the sandy deserts of North Sinai (two reported in 1946 in Harrison and Bates [1991]).

Habits: Nothing known about the Egyptian population. Elsewhere, chiefly hunts by day using a termite mound or similar raised area as a lookout position. Hunts by running the prey down with a burst of speed up to 100kmph sustained for little more than half a kilometer. Diet in Egypt likely to be gazelle, hare, birds, and small mammals. Hunting is by sight. Occupies a home range that varies with food abundance and population density. In Egypt, likely to be very large. Sociability complex. Female generally solitary, except when with cubs or during mating. Young males more sociable, forming bachelor groups. Sociability in Egypt unclear given very low population. Gestation 91–95 days. Cubs in Egypt found in April/May but also November. Calls include a twittering contact call, hisses and snarls when angry, and purrs.

Notes: Hafez (1993) included the Serval *Felis serval* on his *List of the Mammals of Egypt*, but gave the distribution as "Suakin, south of the Sahara." As Suakin is in Sudan, and there are no records from modern Egypt, the species is not included on the Egyptian list.

Similar species: The Leopard is much more thickset with a larger head and spots arranged in distinct rosettes. Within the Cheetah's historical range, there is only one old record from the Western Desert.

LEOPARD *Panthera pardus* (Linnaeus, 1758)
Pl. 14

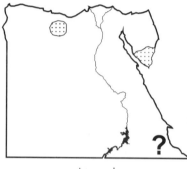

Leopard
(Panthera pardus)

Subspecies occurring in Egypt: *P. p. jarvisi*, *P. p. pardus*.
Arabic: *Nimr*
Identification: Length 170–290cm; Tail 60–100cm; Weight 35–85kg. Male bigger and heavier than female. A large, powerfully-built cat. Orange-buff to pale yellow above, white below. Spotted all over though less densely below and inside the legs. Spots arranged in rosettes (especially on the flanks and hindquarters), the inside of each rosette being slightly darker than the background color. Head large and powerful, spotted more finely than rest of body. Chin white. Ears rather small, marked black and white behind. Tail long, spotted as body for basal half after which the spots amalgamate to form incomplete rings. Underside white. The Sinai Leopard *P. p. jarvisi* has been described as being slightly darker with brownish spots. However, the very few specimens ever collected and available in museums are old and probably faded, thus the spots may be darker in life. Cub, based on one poorly preserved skin, is paler and grayer with longer fur. Otherwise subspecific status based on doubtful cranial characteristics and probably redundant.

Range and status: Most of Africa except for the Cape and the Sahara, though very rare in northwestern Africa. Sinai, southern and western Arabia through Middle East to Turkey. East through Central Asia, Iran, Pakistan, India, Sri Lanka, China, and South Siberia, south to Southeast Asia. In Egypt very rare. One record from the Western Desert (1913) from the Qattara Depression assigned to *P. p. pardus*. Same subspecies also present in the Gebel Elba region where an expedition in 1994 heard a Leopard at night and photographed fresh tracks and spoor the next day. A Leopard was sighted by a Bishari tribesman at a waterhole in 1991, though specific details are lacking. The Sinai Leopard from the mountains of South Sinai is extremely rare. Most recent verified record from the 1950s. Possibly extinct though Bedouin from the Gebel Serbal area and Feiran still claim it is found in the mountains in very small numbers. Listed on CITES Appendix I and classified as vulnerable by the IUCN.

Habitat: Very wide, ranging from deep forest to near desert. In Egypt, the Leopard, with the exception of the Western Desert record, has been found in mountainous areas of Gebel Elba that are lightly wooded with *Acacia* spp.

Habits: Heavily persecuted in Egypt and probably only active at night. Extremely wary and elusive. Diet large to small mammals, birds, reptiles, large insects, eggs, and livestock. The intensively studied Leopard population in the neighboring Negev Desert feeds largely on Ibex *Capra ibex*. Older animals, not able to take Ibex, switch to Rock Hyrax *Procavia capensis*. Hunts by hearing and sight. Needs water but can go a month without. Scent good. Solitary and territorial though nothing known of territory size in Egypt. Likely to be large due to low density of prey species. Nothing known of reproduction in Egypt. Elsewhere, gestation 90–112 days, female giving birth to 1–6 cubs (2–3 norm).

Notes: The Sinai Leopard is a virtually unknown subspecies of the Leopard, its description based largely on skull differences. Because of the large amount of individual variation in the Leopard, it may be

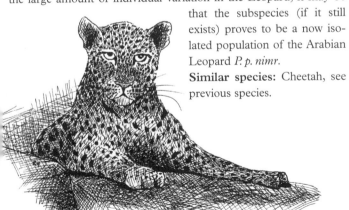

that the subspecies (if it still exists) proves to be a now isolated population of the Arabian Leopard *P. p. nimr*.

Similar species: Cheetah, see previous species.

The Cetaceans—Order Cetacea

There are 76 species of cetaceans worldwide with c. 13 species in Egyptian waters. However, due to the lack of fieldwork, further species may be recorded and certain records, such as that of the Sperm Whale *Physeter macrocephalus*, while not unexpected, have not been fully documented.

With the possible exception of the sirenians, the cetaceans are the mammals most completely adapted to life in the water. Just as the bats' anatomy—internal and external—has evolved for life in the air, that of the cetaceans has evolved for life in the water. While they may bear a superficial resemblance to fish, the cetaceans are true mammals and, like all mammals, have to breathe air. Although some species can dive for long periods and to great depths, they drown if they cannot come up to breathe. Other mammalian features include bearing live young suckled on milk from the female's mammary glands, and, being warm blooded, they control their own body temperature. Though they are hairless as adults (except in the case of certain river dolphins and larger whales), they are never scaled like many fish.

That being said, they are undeniably fish-like in form, with fins instead of limbs, naked, torpedo-shaped bodies, and phenomenal swimming abilities. There are important differences though, which need to be borne in mind, when distinguishing a cetacean from, for instance, a large shark. The key features to look for are the fins and the tail flukes. Cetaceans never have more than three fins, a dorsal fin and two pectoral fins, though in a few species the dorsal fin is absent. Most fish have additional fins and the larger sharks, for instance, have two dorsal and two ventral fins as well as the pectoral fins. All cetaceans have horizontal tail flukes (the power in the water coming from undulating the muscular body up and down) whereas all fish have vertical tail fins, swimming by sweeping the tail from side to side or by using the other fins. At closer range, cetaceans have nostrils, or a nostril, on top of their head known as a blowhole. Fish have external gill flaps or, again in the case of sharks, a series of five to six gill slits behind the eye.

The only other mammal with which the smaller cetaceans could be confused is the Dugong *Dugong dugon* (page 135), which lacks a dorsal fin, has very broad flippers, a very different head shape, and very different behavior.

Given the constraints of their adaptation to their marine, or freshwater, lifestyle, the cetaceans are a diverse group, ranging in size from the Blue Whale *Balaenoptera musculus* that, prior to its decimation by whaling, reached up to 33.5m in length and 190 tons in weight, to the Franciscana *Pontoporia blainvillei* and the Vaquita *Phocoena sinus* that barely reach 1.5m and weigh less than 55kg. Many (including the very largest whales) feed on krill, a tiny marine crustacean that occurs in enormous swarms, while others feed on squid, marine invertebrates, and fish. Only the Orca *Orcinus orca* regularly takes warm-blooded prey such as seals, dolphins, or whales, though Bottle-nosed Dolphins *Tursiops truncatus* have been recorded killing the much smaller Harbour Porpoise *Phocoena phocoena*.

The differences between the whales, dolphins, and porpoises are, for the most part, hazy and not necessarily based on scientific grounds, hence, the preference for the term 'cetacean' for the order. The term 'whale' is generally used in reference to the larger cetaceans, and thus groups the baleen whales, or rorquals, with the totally unrelated Sperm Whale, Narwhal *Monodon monoceros*, Beluga *Delphinapterus leucas,* and beaked whales, as well as the Killer Whale, or Orca, and the pilot whales which are, in fact, dolphins. The term 'dolphin' is generally applied to the smaller cetaceans including the river dolphins, which are largely restricted to freshwater systems and not related to the true dolphins at all. Only the porpoises form a distinct scientific entity being distinguished from the similarly sized, smaller dolphins by the different shape of their teeth. No species of porpoise has been recorded from Egyptian waters.

Of all Egypt's mammals, it is perhaps the cetaceans that are the easiest and most enjoyable to watch. Anyone going out in a dive boat should be rewarded by the sighting of one or more dolphin species, while places such as Ain Sukhna, the islands north of Hurghada, and the Visitor's Center at Ras Muhammad on the Red Sea, and Rasal-Bar on the Mediterranean provide excellent shore-watching sites. The first challenge is to locate the animals and this can best be done by scanning the sea with or without binoculars looking for a telltale dorsal fin breaking the surface of the water. This is clearly much easier on calm, flat days than on choppy days when every white wave crest looks like a breaching dolphin. A second method is to look for secondary signs. It is always worth scanning around fishing vessels, for example, for where there are plenty of fish there is always a heightened probability of dolphins. Similarly, flocks of fish-eating birds, like terns and gulls, can indicate a large fish school and, hence, possible dolphins. To the experienced eye, even without a telltale boat or flock of birds, an area of 'boiling water,' a patch of heavily ruffled water in an otherwise calm sea, can indicate the presence of fish shoals. Of course, the easiest method is simply to wait for the cetaceans to find you. Many species of dolphin bow ride and will readily join boats for periods of a few seconds to fifteen minutes or more. Just leaning over the side of the bows (not too far) can provide fantastic views.

In identifying a cetacean species, certain key identification features should be noted. First comes overall size and shape. Size is best estimated by comparing the length of the animal to the boat, if you are on a boat, but otherwise can be very difficult to estimate. Similarly, shape can be difficult unless the cetacean breaches or a close view can be obtained underwater. Easier to see (and often the first clue as to the presence of a cetacean) is the size, shape, and position of the dorsal fin, if indeed one is present at all. Third, look for coloration and pattern. While most cetaceans are colored rather somberly in blacks, browns, grays, and beiges, they are often boldly patterned. Look for distinctive stripes, spots, and patches, especially around the head and the tip of the beak if there is a beak, and also make note of any visible scarring. The shape of the head is the fourth important point to note. Look for the absence, presence, and size of any beak or melon on the head.

Behavior should also be noted. Look at what the cetaceans are doing and if they exhibit such behavioral traits as lobtailing, breaching, spyhopping, or bow riding. These can be important in distinguishing similar species. The Common Dolphin *Delphinus delphis*, for instance, loves to bow ride while the Risso's Dolphin *Grampus griseus* will do so only rarely. If the cetacean lobtails, make a note of the shape and color of the tail flukes. If it breaches or bow rides then the body coloration can clearly be observed. With the larger whales, the shape of the blow can be diagnostic even at a distance, though care should be taken as a strong breeze can change the shape or angle of a blow. The number of animals seen can also be important though beware: the number of animals visible on the surface at any one time will probably be only a small portion of the total school size.

Finally, location and habitat can be important. Certain Mediterranean species are highly unlikely in the Red Sea and vice versa. Also some species are more likely to be seen in deep water while others favor shallow water and reef margins.

Very little work has been done on the cetaceans of the Red Sea so everyone can contribute greatly to the knowledge of these creatures in the area. Many are known only from a few records and little, if anything, has been looked at in terms of their behavior or distribution in Egyptian waters.

In Arabic, whales are generally known as *hut*, dolphins as *darfil*.

The Baleen Whales—Suborder Mysticeti
10–13 species worldwide with possibly 3–4 in Egyptian waters.

These are the 'Great Whales' along with the Sperm Whale. The suborder includes the four right whales, the Gray Whale *Eschrichtius robustus*, and the rorquals, which are the only group in the suborder to occur in Egyptian waters.

The baleen whales are the largest animals in the sea, including amongst their number the Blue Whale *Balaenoptera musculus*, arguably the largest animal ever to inhabit the earth. Their name comes from the baleen plates that replace teeth in their mouths. These plates are featherlike in form and, in the case of the Bowhead *Balaena mysticetus*, reach up to 4.3m in length. The head and mouth are proportionately large. The whale takes in a vast gulp of seawater that is then forced out of the mouth through the baleen plates by the huge tongue. The baleen filters all the krill, fish, or other food from the water and this is then swallowed. In Egyptian waters, the baleen whales are represented only by the rorquals, the Gray Whale being only found in the North Pacific and the right whales in the northern and southern oceans.

All species of baleen whales, with the exception of the Pygmy Right Whale *Caperea marginata*, were hunted extensively during the days of whaling. It must be stressed that all records of the baleen whales in Egyptian waters are isolated and often old. None are regular in Egyptian waters and with their much-reduced numbers, vagrancy is probably less likely now than in the past.

The Rorquals — Family Balaenopteridae
6–8 species worldwide with 3–4 in Egyptian waters.

The rorquals range in size from the 10m Minke Whale *Balaenoptera acutorostrata*, not yet recorded from Egyptian waters but with isolated records from elsewhere in the Mediterranean, to the Blue Whale described above. All are large, rather streamlined whales with distinctive grooves along the underside of the throat that reach as far back as the navel, except in the Sei Whale *Balaenoptera borealis*.

All rorquals (as predominantly open ocean species) are extremely rare in Egyptian waters owing to its location at the eastern extreme of the Mediterranean and at the head of the Red Sea, both effectively marine dead ends. Indeed, only the Humpback Whale *Megaptera novaeangliae* has been recorded alive here, other records coming from strandings. Of all the rorquals, the Humpback is the most distinctive and easy to identify.

In identifying a rorqual at sea, it is necessary first to distinguish it from the larger, toothed whales. The only toothed whale to approach the rorquals in size is the Sperm Whale *Physeter macrocephalus* that has distinctive wrinkly skin, a hugely bulbous head, no throat grooves, and no dorsal fin. Without exception, all rorquals have a dorsal fin. Rorquals can be distinguished from other larger, toothed whales by the relatively large head and mouth, the splashguard in front of the twin blowholes, the absence of any beak or melon, and a proportionately small dorsal fin.

Distinguishing between the rorqual species can be difficult. Look for the following to aid identification: coloration (uniquely asymmetrical in the Fin Whale *Balaenoptera physalus*); build, size, and position of dorsal fin; size, shape, and color of the pectoral fins and tail flukes; number and extent of throat grooves (if visible); and head shape and presence of ridges on the head. The latter feature is especially important in distinguishing the Sei Whale, recorded from the Gulf of Aqaba, and the Bryde's Whale *Balaenoptera edeni*, as yet unrecorded in the Red Sea but a possible contender for vagrancy. The former has one ridge, the latter three. Finally, the height and shape of the blow can be an important aid to identification, especially at a distance. All records of rorquals in Egyptian waters have been of single individuals, so school size is probably of minimal identification help in the region. Behavior can also be important, the angle of breach being another difference between the Sei and Bryde's Whales.

FIN WHALE (COMMON RORQUAL, FINBACK, FINNER) *Balaenoptera physalus* (Linnaeus, 1758)

Pl. 15

Subspecies occurring in Egypt: unknown but perhaps monotypic.

Identification: Length 18–22m, reportedly up to 25m; Weight 30–80 tons. Male, on average, slightly longer than female. The second largest mammal on earth after the Blue Whale. Huge whale but with very slender build, often described as the 'Greyhound of the Sea.' Body long with

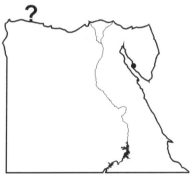

Fin Whale
(Balaenoptera physalus)

small dorsal fin variable in shape and set well back on body. Tail stock thick. Head with single, long ridge running along center of head from blowholes. Coloration unique and diagnostic given a good view. Above pale to dark gray, almost black. Below white, but the under-

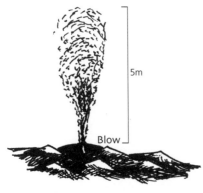

side of the throat is uniquely asymmetrical being dark on the left and white on the right with the coloration often extending onto the lips and also reflected in the baleen. Up to 100 throat grooves. Blowholes prominent. Spout is tall, up to 6m, and slender. Pectoral fins slender, rather short, dark above and pale below. Tail as upper parts above, whitish below, rather pointed with distinct notch in flukes. Rarely appears above surface before diving.

Range and status: Virtually worldwide, from Arctic and Antarctic waters to the tropics where it is less common. From Egyptian waters one record from al-Tor, western Sinai in 1893 and a possible record from Mersa Matruh in 1892. Re-ported as fairly regular in the Mediterranean but most records come from the west. Listed on CITES Appendix I and classified as vulnerable by IUCN.

Habitat: Open ocean. Only coming near land where water is deep.

Habits: Generally found in small groups of up to 5 individuals but schools of up to 100 have been noted in the past. Also solitary. A fast swimmer, it can reach over 30kmph. Dives are erratic and difficult to follow. May dive as deep as 230m. Does not normally show tail flukes before diving. Rarely breaches and when does so, it is at a 45 degree angle, often twisting, and landing with a great crash. Feeds on krill, small fish, squid, etc. Origins of Egyptian specimens unclear due to migratory habits of this species. Spout tall, up to 6m, and relatively slender (see vignette).

Notes: The specimen recorded from Mersa Matruh in 1892 was probably this species but may have been the Blue Whale as efforts to obtain the specimen failed. See Flower (1932).

Similar species: Other large whales, but none have the asymmetrical coloring of the Fin Whale. Blue Whale is longer, much bulkier and differently colored with a spout that can reach 12m. The Sei Whale is smaller with more slender and erect dorsal fin, dark under sides to pectoral fins, indistinct splashguard in front of blowholes and a spout that rarely exceeds 3m. The Humpback Whale has very long, white pectoral fins, knobby appearance, and dark underside and the Sperm Whale has no dorsal fin and much larger head.

SEI WHALE (RUDOLPHI'S RORQUAL) *Balaenoptera borealis* Lesson, 1828
Pl. 15

Subspecies occurring in Egypt: unknown but perhaps monotypic.

Identification: Length 12–16m but reportedly up to 21m; Weight 20–30 tons. Another large whale but considerably smaller than previous species. Upper parts blue-gray to dark gray even blackish. Underparts white with irregular demarcation. Up to 30–60 throat grooves. Dorsal fin slender and rather erect, well back on body. Pectoral fins short and slender, dark above and below. Splashguard in front of blowholes rather low. Single ridge running from guard along top of head. Spout narrow and low, rarely higher

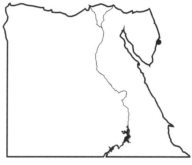

Sei Whale
(*Balaenoptera borealis*)

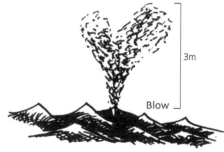

3m

Blow

than 3m. Head symmetrically colored, dark above, light below. Tail stock thick. Tail flukes rather small and broad, with prominent notch in center. Very rarely seen before diving.

Range and status: Virtually worldwide, from Arctic and Antarctic waters to the tropics but less common in the latter where replaced by Bryde's Whale. In Egypt, recorded once from the head of the Gulf of Aqaba, but this record may refer to Eilat, Israel. Recorded from the western Mediterranean. Other regional records from the Gulf of Aden. Listed on CITES Appendix I.

Habitat: Pelagic, though may come inshore around islands.

Habits: Generally in small schools of up to 5 individuals but sometimes in larger congregations. Swims fast—up to 50kmph—when pursued. Dive pattern rather regular compared with other large whales, rarely diving deep or for more than 20 minutes. Tail does not appear above surface before diving. Blows around once a minute when at the surface. Rarely breaches but when it does, breaches at a very shallow angle. Spout fairly low, c. 3m, and narrow.

Notes: Bryde's Whale, the tropical counterpart of the Sei Whale, has yet to be reported in the Egyptian Red Sea but could conceivably occur here. It is very similar to the Sei Whale, perhaps indistinguishable at a distance, but has 3, not 1, ridges running forward from the splashguard, breaches almost vertically, and has a slightly higher spout.

Similar species: Other large whales. For Fin Whale, see above. The Blue Whale is much larger, more mottled with prominent splashguard, tiny dorsal fin, and much taller, more slender spout. The Humpback Whale is all dark with knobby appearance and huge, white pectoral fins. The Sperm Whale has no dorsal fin and a much larger head.

HUMPBACK WHALE (HUMPBACK) *Megaptera novaeangliae* (Borowaski, 1781)

Pl. 15

Humpback Whale
(Megaptera novaeangliae)

Monotypic

Identification: Length 11–16m; Weight 25–30 tons. Large and heavily built but very active whale, with hugely elongated pectoral fins and barely visible dorsal fin. Body large, stocky, black or blackish above, and variable below—sometimes dark, sometimes pale. 12–36 throat grooves, far fewer than in previous two species. Head rather slender with numerous nodules above and below on chin where there may also be white patches formed by colonies of barnacles. Dorsal fin very short and stubby, set well back on body and preceded by a small hump. Tail stock rather narrow and bluntly serrated above. Pectoral fins proportionately very long, up to one-third of total length, with knobby leading edges. May be white on both sides or dark above and white below. Tail flukes very broad with serrated trailing edges and distinct notch. Dark above and normally marked whitish below, the extent and shape of the white areas differing between individuals. Normally shows tail before diving. Spout up to 3m, tall and broad.

Range and status: Widespread but generally migratory. Found in all waters from arctic to tropics. Tends to winter in colder waters and move to tropical oceans to breed. Population from northern Indian Ocean may be sedentary. Two records for the Mediterranean. In Egyptian waters, there is one well-documented example of a young Humpback whale from the Gulf of Aqaba in the northern Red Sea. This whale was seen and photographed off Dahab in 1992 and was a young individual. A dead adult was found off Baltim in 2008. Listed on CITES Appendix I and classified as endangered by IUCN.

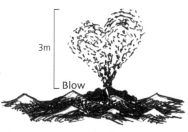

3m
Blow

Habitat: Open oceans but also more inshore, in shallow water, and off islands than other rorquals. The much-studied Hawaiian population, for instance, can be viewed from the land.

Habits: Sociable, in schools of up to 20 animals but generally in smaller groups. For a large whale, extremely active frequently breaching, flipper slapping, spyhopping, and lobtailing—all of which are useful in identification at a distance. Swims slowly and dives are generally shallow, lasting less than 10 minutes. Shows tail prior to diving. When breaches, emerges from water sideways with white of flippers very distinctive. May clear water completely or only half-heartedly. Spout fairly tall, c. 3m, and broad, which is distinctive.

Notes: The Dahab Humpback may have been a genuine vagrant or a stray from the population that is reportedly resident in the northern Indian Ocean.

Similar species: See other large whales, though the huge flippers, broad tail flukes, and, for this region, uniquely knobby head (combined with habits), should make this whale unmistakable.

The Toothed Whales—Suborder Odontoceti
66 species worldwide with 10 in Egyptian waters.

The great majority of the cetaceans belong to the suborder *Odontoceti*, the toothed whales, which include the familiar dolphins, sperm whales, white whales, and porpoises, as well as the much less familiar beaked whales. The term 'toothed whale' is somewhat misleading. While certain dolphin species may have up to 200 teeth, others, such as the Sperm Whale have far fewer and they only emerge in the lower jaw, while the beaked whales, as yet unrecorded in Egyptian waters, generally only have two in the lower jaw and even these do not emerge beyond the gums in the female.

The Sperm Whales—Family Physeteridae
3 species worldwide with 1 possible in Egyptian waters.

The sperm whales are a small family of cetaceans consisting of two small species, the Pygmy Sperm Whale *Kogia breviceps* and the Dwarf Sperm Whale *Kogia simus*, neither of which have been recorded in Egyptian waters, and the Sperm Whale, which is the largest of the toothed whales, reaching over 20m (prior to exploitation by whaling). All species are characterized by the spermaceti organ in the outsized head (which is thought to be related to buoyancy control), the stocky body, and short, rounded pectoral fins, as well as by the undershot lower jaw. The teeth in the upper jaw do not emerge except occasionally in the Dwarf Sperm Whale. For identification of the sole Egyptian species, see below.

SPERM WHALE (CATCHALOT) *Physeter macrocephalus* Linnaeus, 1758
Pl. 15
Monotypic
Identification: Length 11–18m, occasionally up to 21m; Weight 20–50 tons. occasionally up to 70 tons. Males much larger and heavier than females. The largest of the toothed whale and very distinctive with huge head, no dorsal fin, and angled blow. Stout body, dark grayish above, paler, sometimes off-white below. Skin often scarred, especially in old males, and with a wrinkled, prune-like texture. Head huge, up to one-third the total head and body length, with enormous forehead. Eye inconspicuous. No throat grooves. Lower jaw very nar-

row and barely visible when mouth closed. Up to 50 teeth in lower jaw, none in upper. Blowhole uniquely set at one side (the left) and at the front of the head with no splash-guard. Pectoral fins rounded and stubby. No dorsal fin but a series of lumps along the latter half of the back. The first lump is especially conspicuous. Tail stock stout and keeled below. Tail flukes broad with straight trailing edges and notch in the middle. Tail appears above water along with a significant proportion of the body before a dive. Lacks distinctive pale patches on underside except sometimes on underside of head.

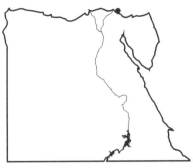

Sperm Whale
(Physeter macrocephalus)

2m

Blow

Blow distinctive, sent forward and to the left due to the site of the blowhole. Can be 2–5m high, but generally 2–3m.

Range and status: Widespread in cold, temperate, and tropical waters but migratory and more frequent in warmer waters in winter, the opposite pattern to the Humpback Whale. In Egypt, one possible record. Teeth (reportedly from this species) were collected at Port Said in 1908–1909 and claimed to be from a specimen stranded along the Mediterranean coast. Has been recorded in the Mediterranean as far east as Greece. Unknown from Egyptian Red Sea. Listed on CITES Appendix I.

Habitat: Pelagic and often in deeper water though has been recorded inshore.

Habits: Very variable. Old males may be solitary while schools of up to hundreds, even thousands, of individuals have been recorded. More normally in schools of 15–20. Thought to be two main forms of social organization: bachelor schools consisting entirely of younger males; and breeding schools of adult females with both male and female young. Generally swims rather slowly. Dives can be extremely deep, to over 3,000m, and it can remain submerged for two hours or more; the fatty tissue in the huge head is thought to act as a buoyancy mechanism. Normal dives shallower but still deep. Shows tail and tail stock before diving. Feeds primarily on squid including the giant squid—the rounded sucker scars of which can be seen on the whale's skin. Often breaches (head shape is diagnostic) and lobtails where shape and tail color should be noted.

Notes: There are recent records from the eastern Mediterranean but not from Egyptian waters.

Similar species: See other large whales. The Short-finned Pilot Whale *Globicephala macrorhynchus* and False Killer Whale *Pseudorca crassidens* both have dorsal fins, much smaller heads, smoother skin, and are much smaller overall.

The White Whales—Family Monodontidae
3 species worldwide with 1 doubtful from Egypt.

The white whales consist of the White Whale or Beluga *Delphinapterus leucas*, the Narwhal, and the Irrawaddy Dolphin *Orcaella brevirostris*. The latter is the only warm water member of the family, found in the seas of Southeast Asia, and it is often included with the true dolphins. The other two species are small whales of the high Arctic. They are characterized by the flexible neck, due to unfused cervical vertebrae, absence of a dorsal fin, and the concave trailing edge to the tail. There is a single record of a Narwhal from Egyptian waters, probably Egypt's most extraordinary mammalian record given the normal range of this very distinctive species. All efforts to find the specimen failed and this 'record' should be treated with skepticism.

NARWHAL (NARWHALE, UNICORN WHALE) *Monodon monoceros* Linnaeus, 1758
Monotypic. Not illustrated.

Identification: Length 3.8–5.5m (not including tusk); Weight 0.8–1.6 tons. A very distinctive whale, particularly due to the male's tusk but also the absence of dorsal fin and unique coloring. Body cylindrical. Upper parts pale grayish, heavily mottled with darker

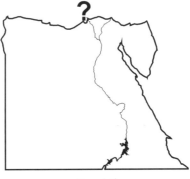

Narwhal
(Monodon monoceros)

gray and inky-black, often darker and more uniform around the face. Underparts white with little or no spotting. Young, uniform gray. Head rather small with small melon and beak virtually absent. Mouth small. Pectoral fins short and dark, often (especially in older animals) with tips curled up. Dorsal fin absent and, at best, only a suggestion of a dorsal hump. Tail stock thick. Tail flukes with a convex trailing edge and, in older animals, the fluke tips may actually point forward. In adult males, there is a long, straight tusk that emerges from the upper lip. This spirals counterclockwise and may be up to 3m in length, though generally smaller. The tusk is present in a very few females (c. 3%), and, exceptionally, a male may develop a second tusk that is invariably shorter than the first. Blow weak but reportedly audible at some distance.

Range and status: A species of the high arctic rarely recorded south of Iceland and very rarely to the British Isles. One record from Holland. In Egyptian waters, reported once and perhaps the most extraordinary of all Egyptian mammal records. In July 1986, a male with a 2m tusk was captured alive by fishermen 500m off the Mediterranean coast at Rosetta and photographed in local papers.

Habitat: Far northern oceans to edge of Arctic icecap. Unconfirmed Egyptian record is only record for the Mediterranean and the most southerly record ever.

Habits: Lives in schools of 1–25 individuals (usually 6–20), generally of a single sex, but occasionally joining in larger congregations. Can swim fast. Dives frequently and deep, to 350m, and does not spend much time at the surface. Food includes squid, crustaceans, small fish, etc. Spyhops and lobtails, but rarely breaches. Tusk is probably of sexual significance as males have been observed fencing with them, the most powerful male achieving dominance and mating more females.

Notes: The unique nature of the Egyptian record would if credible probably relate to an old male caught in adverse currents and/or weather conditions. The specimen is now reportedly in the museum at Rosetta but has yet to be traced.

Similar species: The tusk of the male makes this whale unique. Females or males with broken tusks can be distinguished from all other Egyptian cetaceans by the mottled coloring, and lack of either a dorsal fin or a prominent dorsal hump. Further records of this high arctic species are highly unlikely.

The Dolphins and Blackfish—Family Delphinidae
32 species worldwide with 8 species in Egyptian waters.

The dolphins and blackfish are by far the most likely cetaceans to be encountered in Egyptian waters. The blackfish are sometimes treated separately from the dolphins. They are generally larger than the true dolphins, with blunt snouts, and a more or less developed melon. As the name implies, they are predominantly black, with patches of gray or white in some species, notably the Orca.

The true dolphins are small- to medium-sized cetaceans with a prominent dorsal fin (except in the two extralimital right whale dolphins), a single blowhole placed well forward on the head, a forehead melon, and jaws with many (up to 224) simple, conical teeth. In Risso's Dolphin, there are teeth in the lower jaw only. For the purposes of identification, the dolphins have been divided by color pattern; namely, uniform, patched, and counter shaded, or by the presence or absence of a prominent beak. The same system is followed here. It should be emphasized that this is purely an identification aid and not a scientific classification.

Special note should be taken of dolphin records from Egyptian waters, as a number of species that might be expected to be present have yet to be reported. These include the Striped Dolphin *Stenella coeruleoalba* and the Rough-toothed Dolphin *Steno bredanensis*, both temperate, sub-tropical, and tropical species that have been recorded outside Egyptian waters in the Mediterranean and might also be expected in the Red Sea. The Orca has also been recorded in the Mediterranean and, as one of the most wide-ranging of all cetaceans, could also turn up in Egyptian waters.

COMMON DOLPHIN (ATLANTIC DOLPHIN, SADDLEBACK DOLPHIN)
Delphinus delphis Linnaeus, 1758
Pl. 16

Subspecies occurring in Egypt: uncertain.

Identification: Length 180–260cm; Weight 75–85kg. Male slightly smaller than female. Slender, streamlined dolphin with long, narrow beak and distinct forehead. Dorsal fin varies from erect to slightly sickle-shaped. Pectoral fins are rather long and slender. Coloration variable but distinctive. Above, uniform slate-gray including dorsal fin and (in Red Sea populations at least) flippers. Underside much paler. Diagnostic 'hourglass' pattern along the flanks said to be tan to ocher, but generally appears pale gray in the water. At closer quarters, a dark line running from the flipper to the center of the lower jaw and from the eye to the base of the beak is visible. 33–67 teeth in each side of each jaw (average 40 to 50).

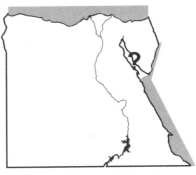

Common Dolphin
(Delphinus delphis)

Range and status: Worldwide in tropical, subtropical, and temperate waters. In Egyptian waters, found in the Mediterranean and the Red Sea extending up the Gulf of Aqaba, possibly Suez, and down to Safaga, Quseir, and Marsa Alam. Status uncertain in the Mediterranean, though there is a record off Port Said and the species is common elsewhere in the Mediterranean. Listed on CITES Appendix II.

Habitat: A pelagic species more likely to be encountered from boats offshore going to or from dive sites rather than by the reefs themselves.

Habits: Forms schools of up to several hundred animals but also in much smaller groups. Swims fast, reportedly up to 40kmph. Joins ships and bow rides but seldom remains for extended periods of time. Crosses and recrosses the path of the boat seemingly effortlessly. Reportedly shy of divers. A good way to locate this species is to watch for flocks of Lesser Crested Terns *Sterna bengalis* or other tern species fishing in large flocks. This indicates a large school of fish and possibly dolphins. Feeds on anchovies, herring, and sardines. Observed feeding on squid at dusk. Gestation 11 months. Calves normally born in spring.

Similar species: Most similar to Spinner Dolphin *Stenella longirostris* from which it can be told by the distinctive hourglass flank pattern. Lack of spots distinguishes it from the Pantropical Spotted Dolphin *Stenella attenuata* and more prominent and sickle-shaped dorsal fin from the Indo-Pacific Humpback Dolphin *Sousa chinensis*. Smaller, less uniform, and with more slender beak than the Bottlenosed Dolphin.

BOTTLE-NOSED DOLPHIN *Tursiops truncatus* (Montague, 1821)
Pl. 16

Two subspecies are probable from Egyptian waters. Mediterranean records probably pertain to the larger *T. t. truncatus* and Red Sea records to the smaller *T. t. aduncus*.

Identification: Length 280–410cm; Weight 150–200kg. A large, uniformly colored dolphin, rather thickset. Dorsal fin slightly sickle-shaped. Dark gray above, paler below without the pattern of the previous species. Beak shorter and stockier, but still distinct with a marked crease where beak meets forehead. 20–26 teeth in each side of each jaw. Lower jaw longer than upper and mouth line curved upward toward eye giving the impression of a 'smile.'

Range and status: Worldwide in coastal-tropical to temperate waters. Recorded from the Egyptian Mediterranean including recent records off Damietta and Port Said. In the Egyptian Red Sea, ranging from near the heads of both the Gulf of Suez and Gulf of Aqaba south to Hurghada, Safaga, and Marsa Alam. Common in the Red Sea and can be seen swimming close offshore at Ain Sukhna and the islands north of Hurghada. A pod is reported resident in a lagoon off Marsa Alam. Listed on CITES Appendix II.

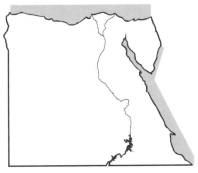

Bottle-nosed Dolphin
(Tursiops truncatus)

Habitat: A coastal species often seen from shore patrolling reef drop-offs and also observed with fleets of local fishermen, e.g., off Ain Sukhna. However, can also be encountered at open sea.

Habits: May form schools of up to several hundred animals but in the Red Sea especially, seems to form smaller pods of 5–10 individuals, especially inshore. These pods seem highly cohesive patrolling the reef drop-offs and several individuals may surface in unison. Groups generally include one or more infant. Swims more leisurely than the previous species, seeming much less urgent or hurried. Less prone to bow ride, though reports from more southern waters show that they do join ships for up to half-an-hour. Gestation 1 year and birth varies from spring to autumn. Feeds on small fish, e.g., mullet, squid, and shrimp. Often breaches. Generally wary of divers.

Similar species: Some individuals may show some spots on the underside but can be told from the Pantropical Spotted Dolphin by much more uniform coloring overall, larger size, and stouter beak without the white tip. The Common, Indo-Pacific Humpback, and Spinner Dolphins are smaller with more slender beaks.

INDO-PACIFIC HUMPBACK DOLPHIN (PLUMBEOUS DOLPHIN) *Sousa chinensis* (Osbeck, 1765)

Pl. 16

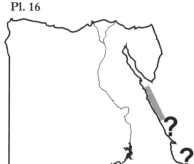

Indo-pacific Humpback Dolphin
(Sousa chinensis)

Subspecies occurring in Egypt: uncertain.

Identification: Length 200–300cm; Weight c. 85kg. Rather small dolphin similar in form to the larger Bottlenosed Dolphin. Uniform dark gray above and along the flanks, paler below. Beak distinct, moderately long and slender. Shows a slight melon on the forehead. Fins distinctive. Dorsal fin only sickle-shaped in young animals; in adults, triangular and low. The humpback in the species name is poorly developed or absent in this species (see notes). The pectoral fins are small and rounded. Fins and beak may be white tipped. Tail stock strongly keeled above and below.

Range and status: Indian and Pacific Oceans from the Red Sea south to South Africa, east to Indonesia. Southern China and northern Australia. In Egypt, recorded in small numbers around the islands at the mouth of the Gulf of Suez (off Tawila, Abu Minqar, Little Giftun, and Gubal Islands). Probably also further south. Listed on CITES Appendix II.

Habitat: Shallow coastal waters and estuaries. Not a pelagic species.

Habits: Solitary, in pairs, or forms small schools of up to 20 animals. Reported to bow ride though shy. Elsewhere, young can be born at any time of the year with a peak in summer. Feeds on reef fish. Sometimes joins schools of Bottle-nosed Dolphins.

Notes: The taxonomy of the Indo-Pacific Humpback Dolphin is complex. Some authorities recognize 3 species *Sousa chinensis, Sousa plumbea,* and *Sousa lentiginosa.* Others treat these as subspecies of one species *Sousa chinensis.* Former system is followed here as this reflects the most current taxonomy.

Similar species: The Bottle-nosed Dolphin is similar in form and color but larger and with very different fins: the dorsal fin is longer, more pointed, and sickle-shaped, and the pectoral fins are more pointed.

PANTROPICAL SPOTTED DOLPHIN (BRIDLED DOLPHIN) *Stenella attenuata* (Gray, 1846)

Pl. 16

Subspecies occurring in Egypt: uncertain.

Identification: Length 210–250cm; Weight to 100kg. Female slightly smaller and lighter than male, on average. Long, slender dolphin with slender beak and distinct forehead. Dorsal fin sickle-shaped and central. Pectoral fins and tail flukes proportionately small and pointed. Coloration variable but nor-

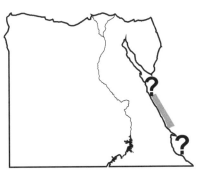

Pantropical Spotted Dolphin
(Stenella attenuata)

mally steel-gray above and white to beige below. Sides of head paler gray than top. Degree of spotting varies between individuals and populations, and also changes over time. Adults spotted, pale gray above, and gray below. Pectoral fins dark gray. Younger animals more uniform but can have a dark stripe running from the pectoral fin to the corner of the mouth, and a black circle around the eye, both of which fade with age. Tip of beak white on both upper and lower lips. Tail stock keeled below and sometimes above. Up to 80 small teeth in each side of each jaw.

Range and status: Tropical Atlantic, Pacific, and Indian Oceans. In Egypt, reportedly off all the main reefs in the northern Red Sea including Shadwan, Tawila, and Giftun Islands, and presumably fur-

ther south. Also unconfirmed off Sinai. Less likely in the Gulf of
Suez. Listed on CITES Appendix II.

Habitat: A pelagic species also found in deeper coastal waters.

Habits: Forms schools of up to several hundred individuals but
generally encountered in smaller groups or pods. Social interactions
complex. Reportedly bow
rides and curious with
snorkelers and divers.
Records suggest
some migration
with most sight-
ings in the northern
Red Sea being between May
and September. Feeds on squid,
small- to medium-sized fish, up to the
size of tuna. At rich fishing grounds, may occur with Risso's Dolphin,
Common Dolphin, False Killer Whales, etc., but such associations yet
to be seen in the Red Sea. Gestation 11 months.

Pantropical Spotted Dolphin
(Stenella attenuata)

Notes: Confirmed sightings (photographs, etc.) still needed though
well within the range and habitat of the Pantropical Spotted Dolphin.
Similar species: No other species of dolphin occurring in Egyptian
waters is as heavily spotted, though the Bottle-nosed Dolphin may
show some spotting on the belly only. Lacks the hourglass pattern of
the Common Dolphin.

SPINNER DOLPHIN *Stenella longirostris* (Gray, 1828)
Pl. 16
Subspecies occurring in Egypt: uncertain.
Identification: Length 180–210cm; Weight 75kg. Small, slender
dolphin with long beak and distinctive habits. Variable in color, pale
gray to nearly black above, lighter on the flanks, often suffused buff
or yellow. Pale to white below and unspotted. Often a pale or dark
line leading from the pectoral fin to the eye. Dorsal fin triangular
and central, often reportedly erect in Egyptian populations (compared by one observer to the dorsal fin of the Orca), though may
also be sickle-shaped. Pectoral fins relatively large, pale to dark gray,
and pointed. Beak long and slender without white tip. Forehead dis-

tinct. Keel above and below tail stock, though variable. 46–65 teeth in each side of upper and lower jaws.

Range and status: Widespread in tropical oceans. Unconfirmed in the Red Sea but likely to occur with reports from the southern Egyptian Red Sea south of Safaga and Marsa Alam. Listed on CITES Appendix II.

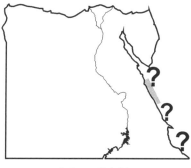

Spinner Dolphin
(Stenella longirostris)

Habitat: Variable. Has been recorded from shallow lagoons, reef drop-offs, and open ocean.

Habits: This species has a complex social life that varies with population and habitat. No work has been done in the Red Sea on this species. Elsewhere, Spinner Dolphins live in schools of 10–100 individuals that spend the day in shallow lagoons resting and socializing. At night, they emerge to hunt in much looser associations. Deep water populations may form schools of up to several thousand individuals. Within these groups, associations of 4–8 individuals may form. Reportedly shy of divers and snorkelers but will bow ride. Has been recorded associating with Pantropical Spotted Dolphins as both prey on tuna (though the latter feeds by day, thus avoiding direct competition). Other food consists of smaller fish and squid. Like the Common Dolphins, schools can be

spotted by disturbance in the water and flocks of seabirds in atten-
dance. The name derives from the species habit of leaping clear o
the water and spinning lengthwise, though the purpose of thi
behavior is unknown. Gestation about 11 months.

Notes: While the Spinner Dolphin has yet to be confirmed in th
Egyptian Red Sea, the habitat is suitable and given the wide range o
the species, its presence is highly probable. The Striped Dolphin i
also possible and has been recorded from the southern Red Sea.

Similar species: The Common Dolphin is probably most simila
but has a shorter beak, less erect dorsal fin, and a distinctive hour
glass pattern along the flanks. The Pantropical Spotted Dolphin i
spotted, and the Bottle-nosed Dolphin is larger and more robust
with a shorter beak and more uniform coloring. The dorsal fin of th
Indo-Pacific Humpback Dolphin is much lower and the pectoral fin
smaller and rounded.

RISSO'S DOLPHIN (GRAMPUS) *Grampus griseus* (G. Cuvier, 1812)
Pl. 16

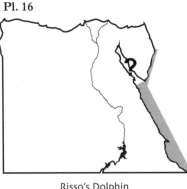

Risso's Dolphin
(Grampus griseus)

Probably monotypic.

Identification: Length 350-
430cm; Weight to 350kg
Large, stocky dolphin nar
rowing considerably at ta
stock. Head rounded an
bulbous with no suggestio
of a beak. Dorsal fin promi
nent, high (higher in males)
and sickle-shaped. Pectora
fins long and tail fins point
ed. Coloration is variabl
but distinct. Basically uni

form gray, paler beneath and white on the chest and chin. Howeve
grows paler with age, especially the head and front edge of dors
fin where it can be almost chalk white. This feature is very notice
able as the animal nears the surface. Males especially are ofte
heavily scarred from rutting, scars showing pale. Teeth only in lowe
jaw, 3–7 on each side.

Range and status: World-wide in tropical, subtropical, and temperate seas. In Egypt, probably relatively common in the Red Sea though no records from the Gulf of Suez. In Gulf of Aqaba, regularly seen off Sharm al-Sheikh and Ras Muhammad. Also recorded off Shadwan Island north of Hurghada. Listed on CITES Appendix II.

Lobtailing

Habitat: An open ocean species, though in the deeper Gulf of Aqaba seen within easy reach of the shore.

Habits: Forms small schools of up to a dozen animals though larger groups have been observed. Does not bow ride but schools may circle a stationary ship seemingly curious. At such times, individuals can be observed lying on their side at the surface of the water presumably to look at the boat. Can be seen breaching and up-ending, exposing only the tail and tail stock above water. Lobtails (see vignette). Pairs seen swimming belly-to-belly and male seen rubbing his penis on a female in precopulatory behavior. Feeds on squid and jellyfish (hence unique dentition). Off Sharm al-Sheikh, observed swimming with Bottle-nosed Dolphins.

Similar species: Lack of beak unique amongst smaller Red Sea dolphins. Distinguished from Short-finned Pilot Whale and False Killer Whale by much smaller size, overall paler coloring, and (in older individuals) by very pale head region, and often extensive scarring.

SHORT-FINNED PILOT WHALE *Globicephala macrorhynchus* Gray, 1846
Pl. 15

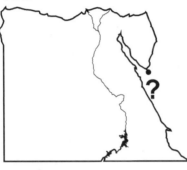

Short-finned Pilot Whale
(Globicephala macrorhynchus)

Probably monotypic.
Identification: Length (male) 475–500cm, (female) 360–425cm; Weight (male) to 2,500kg, (female) to 1,300kg. Large, mostly black dolphin. Torpedo-shaped, rather front-heavy tapering to tail stock. Head with no beak but large, bulbous melon. Upper lip protrudes slightly and mouth curves up in a 'smile.' Mainly black with an anchor-shaped gray patch on the chin and varying amounts of gray on the belly. Dorsal fin strongly sickle-shaped but variable and set rather forward. Pectoral fins moderately long, one-sixth of body length, and sickle-shaped. Tail fins curve back. Tail notch present.

Range and status: Widely distributed in all tropical and subtropical seas. Not found in the Mediterranean. Range and status in Egyptian Red Sea uncertain. Sighted off Ras Muhammad and large schools reportedly present at the mouth of the Gulf of Aqaba in spring/summer 1994. Records from further south near Hurghada may refer to this species or to the False Killer Whale. Listed on CITES Appendix II.

Habitat: The Short-finned Pilot Whale is a deep water species, mainly found offshore.

Habits: Little studied. Off the Canary Isles has been observed in schools of 10 to 30 animals, exceptionally 60. Probably feeds at night on squid, spending much of the day swimming slowly and resting. At rest, lies low in water so that dorsal fin can be difficult to see. Spout small. Shy and sedate. Does not bow ride and rarely breaches or jumps clear of the water. Very tactile, individuals in a group frequently touching each other. Not aggressive, but scarring on individuals suggests some conflicts probably within a group. Schools may be seen lying just below the surface with dorsal fins exposed and all orientated in the same direction—known as logging.

Short-finned Pilot Whale   Logging
(Globicephala macrorhynchus)

Notes: The Long-finned Pilot Whale *Globicephala melas* has been recorded from the Mediterranean but not from Egyptian waters. It is very similar to the current species but has proportionately much longer pectoral fins.

Similar species: For False Killer Whale, see below.

FALSE KILLER WHALE *Pseudorca crassidens* (Owen, 1846)
Pl. 15

Probably monotypic.

Identification: Length (male) 425–600cm, (female) 365–520cm; Weight (male) to 2,000kg, (female) to 1,200kg. Another large, black dolphin. Long and torpedo-shaped, less front-heavy than previous species. Uniform glossy-black sometimes paler below. Upper jaw overlaps lower. Head is slender and tapered without the pro-

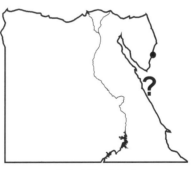

False Killer Whale
(Pseudorca crassidens)

nounced melon of the previous species. Dorsal fin relatively large and sickle-shaped, set well back. Pectoral fins rather short, slightly rounded at tip, and distinctive for the hump on their leading edge.

Range and status: Worldwide in tropical to temperate seas including the Mediterranean though not recorded in Egyptian waters. Status in Egyptian Red Sea uncertain due to confusion with the previous species. Edwards and Head (1987) note two individuals from the Gulf of Aqaba. However, the habitat is suitable and it ranges throughout the Indian Ocean. Listed on CITES Appendix II.

Habitat: Pelagic, rarely found in shallow waters.

Habits: Little known due to their open water habitat. Highly sociable, occurring in schools of several hundred, even thousand, individuals. Feed predominantly on squid and small- to medium-sized fish but have been known to attack other dolphins and even the calves of Humpback Whales.

Similar species: The Short-finned Pilot Whale is superficially very similar. However, it has a clear melon, longer pectoral fins (that lack the hump on the leading edge), and the dorsal fin is set further forward. The Short-finned Pilot Whale occurs in smaller schools. The Orca, another possible species, is larger with a distinct saddle, white patch above and behind the eye, and white underparts.

The Sirenians — Order Sirenia

L arge marine and freshwater herbivores. Entirely aquatic with no external hind limbs and forelimbs adapted as flippers. Tail either rounded or fin-like. Confined to tropical and subtropical waters, with the exception of the extinct Steller's Sea Cow *Hydrodamalis gigas*. Distinguished from most cetaceans by the lack of a dorsal fin. There are two families, the manatees (Trichechidae) and the Dugong (Dugongidae), of which only the latter is represented in Egypt. In Arabic, it is often known as *'Arus al-bahr*.

The Dugong — Family Dugongidae

Single species, the Dugong *Dugong dugon*, for description of which see below.

1 species with 1 found in Egypt.

DUGONG *Dugong dugon* (Miller, 1776)

Pl. 17

Monotypic

Identification: Length 2.5–3.2m but up to 4m recorded; Weight 150–500kg, but up to 1,012kg recorded. Male larger and heavier than female. Entirely aquatic mammal unrelated and distinct from the whales and dolphins. Large, cigar-shaped body, gray-brown above, paler below, appears hairless. Forelimbs as rounded flippers without nails. Hind limbs absent. Head with blunt snout, curved downward and with bristles on margins of upper lip. Nostrils on top of snout but not behind head as in dolphins. Eyes small. No external ears. Tail like dolphins with flukes, flattened horizontally.

Range and status: Red Sea and coast of East Africa south to

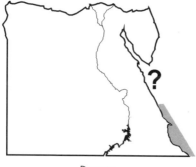

Dugong
(*Dugong dugon*)

Mozambique, Madagascar, east to India and Sri Lanka, Indonesia, Philippines north to Taiwan and south to New Guinea and Australia. In Egypt, in the southern Red Sea where, at best, very rare. One unconfirmed record from the early 1980s from the eastern side of Tiran Island. Most recent reports from 1994 from south of Marsa al-Alam. Possible sighting in 1994 just north of Abu Minqar Island, Hurghada should be treated with caution. Previously more common with photographs and specimens of Dugongs from the Egyptian Red Sea at the Oceanographic Institute in Hurghada. Disturbance and habitat destruction (with the boom in tourist development) and drowning in fishing nets have probably caused their decline rather than direct hunting. Despite rarity, not a protected species in Egypt. Listed on CITES Appendix I/II and classified by the IUCN as vulnerable.

Habitat: Shallow, inshore water and lagoons with underwater vegetation.

Habits: Unknown in Egypt. Elsewhere, spends day in deeper water moving inshore at night to browse. Generally seen singly or in pairs or small groups, occasionally up to 100 animals, though this is highly unlikely in Egyptian waters. Vegetarian, feeding entirely on underwater weeds and especially sea grasses. Swims slowly, even when escaping danger. Dives for up to 5 minutes but normally for 1–2 minutes. No known predators apart from humans, though possibly large sharks. Mating front-to-front, vertically in water, heads above surface. Gestation probably around 11 months. Single young. No known breeding season.

Notes: The only other sea mammal recorded from Egyptian waters, aside from the cetaceans, is the Mediterranean Monk Seal *Monachus monachus* found only on the Mediterranean coast and extinct now in Egyptian waters, though possibly still found in Libyan waters.

Several Mediterranean Monk Seals were killed in World War I in Egypt but it clung on until 1919 and by 1920, only one individual remained at Sallum, with specific orders from the army not to be killed. This was the last Egyptian record.

Similar species: Told from all Egyptian dolphin and whale species by absence of dorsal fin and nostrils in front of crown on downward-pointing muzzle. Also much slower moving. Only record of a finless cetacean in Egyptian waters is a possible (improbable) Narwhal in

1995, but this was in the Mediterranean where the Dugong is long extinct. Biggest confusion may arise from a surfacing sea turtle, especially the Leatherback *Dermochelys coriacea*, which may reach 2.9m in length. However, this species has a distinctly ridged back, and all sea turtles lack the tapered body and forked tail.

The Odd-toed Ungulates — Order Perissodactyla

A small order of herbivorous mammals, formerly far more wide-spread, of which three families survive today: the horses and donkeys (Equidae), the rhinoceroses (Rhinocerotidae), and the tapirs (Tapiridae). All are characterized by an odd number of toes, one in the equids and three in the other two families. Formerly, the Perissodactyla were far more widespread and an additional three families are known only from the fossil record.

The Horses and Asses — Family Equidae

This family is characterized by the reduction of the digits on each foot to one, namely, the hoof. The basic form of the horse and donkey is too well-known to warrant elaboration.

7 species of which 1 is found in Egypt (in the wild).

AFRICAN WILD ASS (AFRICAN ASS, WILD ASS) *Equus africanus* Fitzinger, 1857

Pl. 19

Subspecies occurring in Egypt: *E. a. africinus.*

Arabic: *Humar barri nubi*

Identification: Length 235–250cm; Tail 40–45cm; Shoulder 115–125cm; Weight c. 275kg. Typical donkey in form with short, smooth coat. Above and flanks, pale gray-tinged ruddy, below white. Legs paler than body and without stripes. Dark stripe along back of spine from shoulder to base of tail, with a transverse stripe across the shoulder generally present. Head large with white ring around eye, white muzzle and jaw. Ears large and elongated, inside white, back grayish. Short, erect mane from between ears down neck to base of neck. No forelock. Tail long, with blackish tail tuft. Voice as for domestic donkey.

Range and status: Formerly much of North Africa from Morocco west, south to the Horn of Africa and north to Syria. Now, much reduced and confined to coastal ranges of Red Sea, mainly in Sudan, and parts of Somalia, northeastern Ethiopia, and possibly

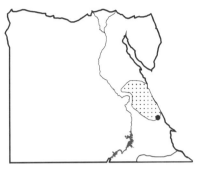

African Wild Ass
(Equus africanus)

Eritrea. In Egypt, status uncertain and may be extinct. Last sightings in Gebel Elba region and head of Wadi Allaqi. Recent possible records from tracks and dung at Bir Kimotit, Wadi Mitkwan, and Wadi Akau in the southeast. These could easily be domestic donkeys, but local Bishari tribesmen do differentiate between wild and domestic animals. At best, very rare and threatened—if not by direct hunting then by hybridization with domestic donkeys. May also cling on along Libyan–Egyptian border north of Siwa, where subspecies reported to have a triple shoulder cross. Protected by Egyptian law and under CITES Appendix I. Listed as endangered by IUCN.

Habitat: Plains and wadis in desert or semi-desert regions. Good climber and can inhabit more rugged regions. Can go periods without water, but needs to drink probably at least every 3 days.

Habits: Unknown in Egypt. Elsewhere, generally active at night or early morning and evening. Rests during the day. Male may be solitary or go around in bachelor herds. Females gather in herds, sometimes with young. Sub-adult male and female also in herds. Herd size may be 6–12 or as high as 50 (highly unlikely now). Does not form stable groups and all herds are temporary. Diet consists of grass and other plants, rarely leaves. Only recorded predator in Egypt is the Leopard *Panthera pardus*, which is itself, at best, very rare. Gestation 330–365 days. Single foal is the norm.

Notes: Beware of untended groups of donkeys in the southern Eastern Desert. There is a tendency for these animals to breed back to type. See below.

Similar species: In Egypt, can only be confused with the domestic donkey from which the African Wild Ass can be told by paler coloring (with legs paler than body), the lack of stripes on the legs, and lack of dark tip to the ears. In the past, the Bishari would let their female donkeys loose in the desert to be mated by wild stallions. Any sightings of the African Wild Ass should be made with the utmost care due to the presence of domestic animals and the danger of hybrids. If an adult animal can be approached at all closely, it is virtually certain not to be wild.

The Hyraxes—Order Hyracoidea

See below for a description of this single family order.

The Hyraxes—Family Procaviidae

The hyraxes are a peculiar family. In general form, they look very much like large rodents—indeed, their family name means 'before the guinea pigs.' Their anatomy, particularly the foot structure, seemed to indicate a common relationship with the elephants and the sirenians but recent studies have demonstrated a closer relationship with the odd-toed ungulates. Whatever the relationship, the hyraxes as a group go back a long way. In the desert near Fayoum, fossils have been found that indicate that the ancestors of the modern hyraxes, many much larger than the present day species, were the dominant herbivores in the region some 40 million years ago. Today, the hyraxes are hare-sized mammals with short ears, a coarse coat, and no visible tail. The dentition is notable for the greatly enlarged incisors. Other characteristics include a limited ability to internally control body temperature.

The hyraxes are generally split into three groups, the rock hyraxes *Procavia*, bush hyraxes *Heterohyrax*, and tree hyraxes *Dendrohyrax*. All species are characterized by a series of skin glands that extend down the center of the back and are often surrounded by an erectile mane of hairs of distinctive color.

11(?) species with 2(?) found in Egypt. Both are known in Arabic as *wabar*.

ROCK HYRAX (CAPE ROCK HYRAX, CAPE DASSIE, DASSIE, CONEY)
Procavia capensis (Pallas, 1766)
Pl. 18

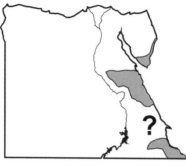

Rock Hyrax
(*Procavia capensis*)
and Yellow-spotted Hyrax
(*Heterohyrax brucei*)

Subspecies occurring in Egypt:
P. c. syriaca and *P. c. ruficeps*.
Identification: Length 345–574mm; Tail absent; Weight 2.5–5.0kg. Stocky, rabbit-like mammal with short ears and no external tail. Upper parts gray through gray-brown to ocher, flanks and underparts, including inside of limbs and throat, paler and buffish. Large glands on back sometimes surrounded by more yellow-orange hair, but this feature is variable. Coat short and dense, liberally scattered with much longer, touch-sensitive hairs. Feet distinctive: forefoot with four very short toes; hind feet with three, one of which is elongated, separated from the others with a curious, curved claw, and used as a grasping toe. Neck very short giving the impression that the head is just stuck straight onto the body. Snout pointed but short. Nasal region naked and black. Whiskers well-developed and black. Eyes rather large and dark with pale patch often visible above eye. Ears small and rounded with paler patch behind base. External tail absent.

Range and status: Throughout Africa south of the Sahara, except for rainforest regions. Isolated populations in North Africa, but range continuous from Sudan north through Egypt to Sinai and on to Israel, Jordan, Lebanon, and Arabia. In Egypt, *P. c. syriaca* is found in South Sinai north to Taba (along airport road), including St. Katherine, Wadi Kid (skull), Ras Muhammad, and al-Tor. *P. c. ruficeps* recorded from the Eastern Desert south of a line running roughly from Safaga to Sohag and including Gebel Elba region, Bir Kimotit and Wadi Nagib, and Wadi Seiga, west of Wadi Allaqi. Historically, recorded further north to Wadi Gattaar, west of

Hurghada. However, some of these sightings may be of the Yellow-spotted Hyrax *Heterohyrax brucei* and the range map combines records of both species for this reason. Everywhere much reduced due to hunting and clearance of acacia trees for charcoal, robbing the Rock Hyrax of its main food source.

Habitat: Entirely restricted to rocky areas with cliffs and availability of acacia. Dens apparent by the streaking of white urine, but as this can stay around for many years, such streaking does not mean a den is active.

Habits: Mainly active by day with apparent peaks in early morning and evening. Lives in colonies in rock crevices and dens on often inaccessible cliffs. One such colony at Bir Kimotit consisted of 7 adults and 3 young but can be much larger. Very alert with at least some members of the colony acting as lookouts and alerting others to danger with sharp alarm call. Adept climber, foot structure allowing purchase on the smoothest of rocks. Sight excellent, hearing and scent good. Diet largely vegetarian and predominantly acacia leaves and seeds, though may take a wide range of other food including insects and small reptiles. Does not need to drink water. Predators include birds of prey (range of Verreaux's Eagle *Aquila verreauxii* almost exactly mirrors that of the Rock Hyrax in Africa, but it is very rare in Egypt), Jackals *Canis aureus*, Wild Cats *Felis silvestris*, and Leopards *Panthera pardus*. In Israel, Leopards have been reported to turn to hyraxes when they were too old or weak to tackle ibex. Mating in August/September. Males establish and defend territories. Mating brief and vigorous. After mating season, individuals mix freely again. Gestation 7.5–8 months. Litter size 1–6.

Notes: For the intriguing role of the hyrax in Bedouin lore, refer to Hobbs (1990). The taxonomy of *Procavia capensis* is complex. Kingdon (1997) splits the rock hyraxes into five species, naming the Egyptian representative *Procavia habessinica* with a range through northeastern Africa and Arabia.

Similar species: For Yellow-spotted Hyrax, see below. The Cape Hare *Lepus capensis* has much larger and longer ears and is found in a completely different habitat.

YELLOW-SPOTTED HYRAX (YELLOW-SPOTTED ROCK HYRAX)
Heterohyrax brucei (Gray, 1868)
Pl. 19
Subspecies occurring in Egypt: uncertain but possibly *H. b. thomasi*
and *H. b. hoogstraali.*

Identification: Length 325–560mm; Tail absent externally; Weight
1.3–3.6kg. Similar to the Rock Hyrax, it is stocky and rabbit-like with
no external tail. Ears small and rounded. However, generally smaller
and slighter with a slightly more slender muzzle. Pelage rather wool-
ly. Upper parts and flanks generally grayish, especially in arid areas,
with a salt-and-pepper appearance. Guard hairs long and black
tipped. Dorsal gland surrounded by erectile hair and frequently
forming a yellow patch, hence, the English name. Underparts clearly
demarcated, ranging from white to cream. Pale to white eyebrows
very distinctive, even from a distance. Foot structure as for previous
species. Vibrissae long and abundant.

Range and status: Recorded from much of southern, eastern, and
northeastern Africa to Sinai though status uncertain and no Egyptian
specimens examined. Not mentioned from Egypt by Osborn and
Helmy (1980), or from Sinai by Harrison and Bates (1991).
According to Kingdon (1997), present along Red Sea hills and Sinai
though no specific locations given. Distribution map is of both
hyraxes (see note on previous species).

Habitat: Similar to previous species though perhaps less tied to
rocky habitat. Elsewhere in range has been found in more open habitats using abandoned burrows of other animals rather than rock dens.

Habits: Much as previous species. Mainly active by day with two
main periods of activity in early mid-morning and late afternoon.
Avoids the heat of the day though does bask. Lives in colonies with
numbers up to 34 recorded but usually smaller. Territorial males
found with a variable number of adult females as well as sub-adult
and juvenile males and females. Adept climber, not just of rocks but
also of trees. In parts of range, *Acacia* spp. make up a large part of the
diet though in other areas has a more catholic diet. Does not need to
drink water. Dust bathes. Predators as for previous species.
Advertising call distinctive (see below).

Notes: The Yellow-spotted Hyrax and the Rock Hyrax often occur in mixed colonies and freely interact with each other though this habit not confirmed in Egypt. As with *Procavia*, the taxonomy of *Heterohyrax* is complex with most authors recognizing one extant species while others, such as Kingdon (1997), recognize as many as three.

Similar species: Rock Hyrax is larger and more robust, with a browner and more uniform coat, and buffer, less differentiated underparts. Pale, less prominent eyebrows. Can also be differentiated by call. According to Kingdon (1997), the call of the Rock Hyrax is a "long series of ascending yelps descending to grunts," and that of the Yellow-spotted Hyrax a "penetrating series of whining mews."

The Even-toed Ungulates—Order Artiodactyla

A large order of some 187 species that includes many familiar families such as the pigs (Suidae), cattle and allies (Bovidae) deer (Cervidae), and camels (Camelidae), as well as the less familiar peccaries (Tayassuidae) and chevrotains (Tragulidae). They range worldwide from the Musk Ox *Ovibos moschatus* and Reindeer *Rangifer tarandus* of the high arctic to the Hippopotamus *Hippopotamus amphibious* of the steamy tropics, the Yak *Bos mutus* of the towering Himalayas to the Dromedary *Camelus dromedarius* of the searing desert. This diverse array of mammals is connected by certain anatomical features, not least, an even number of toes on each foot, four in the pigs and allies, suborder Suina, and the ruminants suborders Tylopoda and Ruminantia.

Today, the Artiodactyla are represented in Egypt by only one family, the Bovidae, the cattle and antelopes together with the extralimital Pronghorn *Antilocapra americana* and the duikers *Cephalophinae*. This was not always the case, however. The family Hippopotamidae, the hippopotamuses were represented here by the Hippopotamus *Hippopotamus amphibious*, possibly until the 1800s. The pigs (Suidae), survived until 1912 when the last Egyptian Wild Boar *Sus scrofa* died ignominiously in Giza Zoo. The One-humped Camel or Dromedary of the Camelidae is still wide spread as a domestic species but, as elsewhere, extinct in the wild Sadly, most of the even-toed ungulates recorded from modern Egypt are now extinct, almost extinct, or, at best, much reduced and rare.

The Bovids—Family Bovidae

The bovids include a diverse array of species represented in Egypt by mammals as diverse as the stocky Nubian Ibex *Capra ibex* and the slender and diminutive gazelles *Gazella* spp. All are specialized herbivores with a multi-chambered stomach and prominently ridged cheek teeth. All have two functional toes. Several species are now doubtful from Egypt and at least two, the Arabian Oryx *Oryx leucoryx* and the Bubal Hartebeeste *Alcelaphus buselaphus buselaphus*, are locally extinct.

ADDAX *Addax nasomaculatus* (de Blainville, 1816)

Pl. 19

Monotypic

Arabic: *Baqar wahshi*

Identification: No measurements from Egypt. Length 145–210cm; Tail 25–35cm; Shoulder 95–115cm; Weight 60–135kg. Male slightly larger and heavier than female. Size of a lightly built donkey, but on much more graceful lines. Upper parts pale graybrown, paler with age and paler in winter than summer,

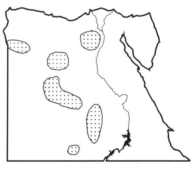

Addax
(Addax nasomaculatus)

also with individual variation. Rump, underparts, and legs whitish. Dark brown patch on crown in front of horns. Calves more reddish with darker crown. Feet with large, flat hooves. Head with rather short muzzle, whitish around lips and with distinct white band across the muzzle below the eyes. Ears not especially large, white behind. Both sexes with long, slender, spiraled horns up to 89cm long along the spiral in the male. Ringed for first two-thirds. Horns of young animals weakly spiraled or straight depending on age. Tail extends to hock, whitish with white terminal tuft. Voice little known, but grunting recorded.

Range and status: Formerly much of North Africa from southern Mauritania and Senegal in the west across to Egypt and Sudan in the

Addax
(*Addax nasomaculatus*)

east. Today, population has been decimated by hunting and now only occurrs in a few localities in Niger, Mali, Chad, and Mauritania. Unless rigorously protected, could become extinct in the wild. In Egypt, formerly over much of the Western Desert east to the margins of the Fayoum and around all the major Western Desert oases. In pharaonic times, possibly also in the Eastern Desert. Population now decimated and almost certainly extinct in Egypt, the country with the last report of a live animal, according to some authorities, being from 1900 west of Alexandria and 1931 near Bir al-Shab west of what is now Lake Nasser. Old skulls and horns continued to be found into the 1970s. If the Addax still exists in Egypt, it would only be in very small numbers in the remotest parts of the desert, but this is highly unlikely for this globally threatened animal. Listed on CITES Appendix I and classified by the IUCN as endangered.

Habitat: The Addax is perhaps the antelope species most adapted to living in the desert where it can thrive in sandy or *hamada* desert in the absence of standing water.

Habits: In Egypt unknown. Elsewhere, most active in coolest part of the day and during the night, resting during the heat of the day. Not territorial and nomadic, formerly in large herds of over a thousand now in much smaller groups of up to 20 led by an old male. Can travel great distances in search of food, which it probably locates by scent. Vegetarian diet, feeding on grass, leaves, etc. Does not need standing water, able to obtain sufficient moisture from plants and from dew, but will drink readily if water is available. Sight, scent, and hearing all acute. Very wary. Predators, such as the Leopard *Panthera pardus* and Striped Hyena *Hyaena hyaena*, largely exterminated from known range in Egypt. Decline entirely due to hunting by humans. Heavy build (for an antelope) means the Addax is not a fast runner making it easy game for modern hunters. Gestation 257–264 days giving birth to 1 young in early spring.

Notes: Interesting references to the Addax in the region can be found in Ralph Bagnold's *Libyan Sands—Travel in a Dead World* (1987 ed.).

Similar species: The only Egyptian antelope with spiral horns. Scimitar-horned Oryx *Oryx dammah*, also almost certainly extinct in Egypt, has swept-back, curved horns, and reddish patches on neck and hindquarters.

SCIMITAR-HORNED ORYX (SCIMITAR ORYX, NORTH AFRICAN ORYX, WHITE ORYX) *Oryx dammah* (Cretzschmar, 1826)
Pl. 19

Monotypic but see note.
Arabic: *Maha, Abu hirab*
Identification: No measurements from Egypt. Length
235–280cm; Tail. 45–60cm;
Shoulder 110–125cm; Weight
135–180kg. Large, pale antelope, distinctively marked
and with long, back-curved
horns. Pale, almost white
above and below with indistinct or no flank stripe. Pale
chestnut to reddish on neck,

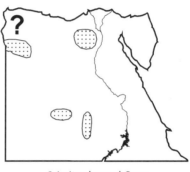

Scimitar-horned Oryx
(Oryx dammah)

shoulders, along back to tail tuft. Legs white. Head whitish with brown along top of snout and between horns and patch or line running vertically through the eye. Horns very long and curved back, thicker in male, thinner and slightly longer in female. Ringed. Tail down to below hock with terminal tuft. Voice a low grunt or, when threatened, a bleat.

Range and status: Formerly over much of North Africa from Senegal north to southern Morocco east through Niger, Mali, Chad to Libya, Egypt, and Sudan, east to Sinai. Now, like the Addax, exter-minated over much of its range, largely due to hunting, and in dan-ger of extinction in the wild. Captive breeding program currently underway in Tunisia. In Egypt, formerly in Western Desert recorded in nineteenth century from around Siwa, Kharga, Dakhla, Wadi

Scimitar-horned Oryx
(Oryx dammah)

Natrun, and the Fayoum. Now almost certainly extinct in the country although there is one sight record by Osborn and Helmy (1980) from south of Mersa Matruh in 1972. Listed in CITES under Appendix I and classified by IUCN as endangered.

Habitat: Arid desert to semi-desert and steppe habitat. Not limited by availability of standing water.

Habits: Unknown in Egypt but elsewhere much as Addax. Active in early morning and early evening and sometimes during night. Rests during day in shade, if available. Sociable, living in small herds of up to 20–30 animals led by an old male. In past, much larger groups recorded. Old males may be solitary. Nomadic, following rains to areas of vegetation. Diet includes grass and leaves, also roots and fruits, such as wild melons. Can survive on moisture from food and dew in absence of standing water. If water is available, will drink often. Predators such as Leopard and Hyena largely exterminated in modern Egypt where main threat, if it still survives, is human. Gestation 242–256 days. Gives birth to a single calf but timing of birth unknown in Egypt and variable elsewhere. Lives up to 22 years in captivity.

Notes: The Scimitar-horned Oryx is treated here as a full species but some authors treat it as a subspecies of the Oryx *Oryx gazella*, namely *Oryx gazella dammah*.

Similar species: If still present in Egypt, it could only be confused with the Addax, which has no reddish patch on neck, spiral rather than curved-back horns, and is dirty-gray. Horn shape should distinguish from similarly colored cattle that would be highly unlikely in areas still possibly inhabited by the Scimitar-horned Oryx.

DORCAS GAZELLE (JEBEER) *Gazella dorcas* (Linnaeus, 1758)
Pl. 20

Subspecies occurring in Egypt:
G. d. dorcas, G. d. saudiya, G. d. Isabella, and *G. d. littoralis.*
Arabic: *Ghazal, 'Afri*
Identification: Length (Male) 98–117cm; Tail 10.2–18.0cm; Shoulder 55–65cm; Weight 14–20kg. Stocky, rather thickset gazelle with slender limbs and striking head pattern. Males generally slightly larger than females, though horn size and shape is

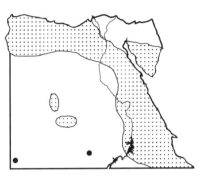

Dorcas Gazelle
(Gazella dorcas)

best distinguishing feature. Upper parts pale red-brown, darker in winter, with indistinct darker flank stripe absent in Dorcas, and white underparts. Legs as upper parts outside, inner side white. Patch of dark hair immediately above margin of hoof and black stripe running up back of lower leg from hoof (not obvious in field). Hoof dark. Head with short muzzle and rather thick neck. Strongly marked facial region with dark central strip from between horns to nasal region. This is bordered by a white stripe running from below the horns above and around the eye again down to the nasal region. Below this, running from the eye, is a further dark facial stripe. Cheeks as upper parts. Throat white. Ears very large and pale, buffish on inside. Horns in both sexes. In male, thick and lyre-shaped, curving back and then up and may or may not be curved in at tip, except in *G. d. saudiya* where almost straight and often only slightly curved in at tip. Strongly and broadly ringed, though last quarter smooth. In female, horns are much more slender, straighter, and less strongly ringed. Tail dark on upper side contrasting with white of rump area. Generally silent but has a short bark when alarmed and longer bleat when threatened.

Range and status: Formerly much of North Africa south to Senegal and north to Morocco in the west, east across Algeria, Tunisia, Libya, Chad, Egypt, and Sudan to Ethiopia. Also Sinai, parts of Arabia, and

Near East. Range in east unclear due to taxonomic confusion with other gazelle species. Over much of its range, it is now much reduced because of hunting. In Egypt, formerly widespread in both Western and Eastern Desert, and Sinai. Now, although the most likely gazelle to be seen in Egypt, much reduced due to indiscriminate hunting and, for this reason, details of recent sightings cannot be published here. *G. d. dorcas* recorded from Western Desert, *G. d. littoralis* from Eastern Desert, and *G. d. saudiya* from Sinai. Some authors refer specimens from North Sinai to *G. d.dorcas* and those from the South to *G. d. isabella*. Found in a number of Protected Areas including St. Katherine. Protected by Egyptian Law, though law simply not enforced. Listed by IUCN as insufficiently known.

Habitat: Stony desert and coastal plains with vegetation. In Eastern Desert and Sinai, vegetated wadis. Also margins of sandy desert.

Habits: Mainly active in early morning and late afternoon, but may also feed during the day and, in areas where persecuted, has been recorded as nocturnal. Mainly found in pairs but also in small herds of up to 30–40 or larger groups on migration, though such congregations probably no longer occur. Very wary with excellent senses of sight, smell, and hearing. Males territorial during breeding season and may hold a territory with one or more female. Diet is vegetarian taking a wide variety of grasses and plant species including *Acacia* spp., also seed pods and fruits, and, in the past, crops. Requires standing water and cannot survive solely on moisture from vegetation during dry periods. Predators such as Leopard and Striped Hyena have been largely exterminated though the Cheetah *Acinonyx jubatus* may still survive in Qattara Depression. Main threat is indiscriminate and uncontrolled hunting by humans, especially with increase in use of high-power rifles and 4-wheel drives. When alarmed, utters short bark and then runs off with a distinct, stiff-legged gait with tail raised. Gestation 169–181 days. Births in Egypt recorded from February to April and September to October. One calf per litter.

Notes: The taxonomy of gazelles is frequently changing and a detailed discussion of it is beyond the scope of a field guide. The three main subspecies here are taken from Osborn and Helmy (1980). The distribution of the fourth subspecies, *G. d. isabella*, and *G. d. dorcas* in Sinai is proposed by Ferguson (1981).

Similar species: The Slender-horned Gazelle *Gazella leptoceros* is much paler with more slender horns and confined to a small area of the Western Desert, if indeed it still survives in Egypt at all. The Mountain Gazelle *Gazella gazella* is larger, longer-limbed, with shorter horns and ears, and slightly less clearly defined facial pattern. It is only known from one sight record in Egypt from the 1930s and should only cause confusion along the Egyptian–Israeli border.

SLENDER-HORNED GAZELLE (LODER'S GAZELLE, WHITE GAZELLE, RHIM) *Gazella leptoceros* (F. Cuvier 1842)
Pl. 20

Probably monotypic, though note comments on gazelle taxonomy for previous species. Osborn and Helmy (1980) give subspecies found in Egypt as *G. l. leptoceros*.
Arabic: *Rim, Ghazal abyad*
Identification: Length 101–116.5cm; Tail 12.5–16.6cm; Shoulder 65–72cm; Weight 15–30kg. Male, on average, larger and heavier than female. Upper parts

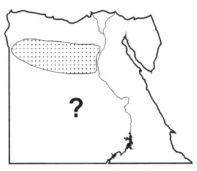

Slender-horned Gazelle
(Gazella leptoceros)

pale buff to cream, flanks with lighter stripe and then darker one, demarcating white of underside. Limbs very pale to white, color of upper parts not extending onto them. Face much paler than the Dorcas Gazelle with much less distinct facial markings and no brown or red-brown. Ears very large, elongated, and pale. Horns long (up to 41cm in male, shorter in female) and only slightly curved, never as strongly lyre-shaped as in Dorcas Gazelle, and thinner with prominent rings but smooth tips that can occupy as much as half of the horn's length. Female horns more slender still, and straighter. Tail blackish brown, very distinct against pale rump.
Range and status: Formerly over much of North Africa from south of the Atlas Mountains to central Algeria, Tunisia, Libya, and Chad to Egypt and possibly Sudan. Now exterminated from much of this

range. In Egypt, formerly in northern Western Desert from Siwa around the margins of the Qattara Depression to Wadi Natrun and south to the Fayoum. Now limited to an area southwest of the Fayoum, if indeed it is not extinct in Egypt. If it survives, the Egyptian population could make up a significant proportion of the remaining global stock. Protected by Egyptian law, though this is largely unenforced. Listed as endangered by the IUCN.

Habitat: Generally a gazelle of sandy desert and dune areas with sufficient vegetation. Also found in more rocky areas. Absent from Mediterranean coastal desert where Dorcas Gazelle is, or was, present. More of a desert species than the previous species and reported to be able to survive without standing water, but will drink if water is available.

Habits: Little known in Egypt, or indeed elsewhere. Like most large desert herbivores, most active during cooler parts of the day and possibly at night, resting up in shade of *Acacia* spp. or shrub during hotter parts of the day. Sociability little known but probably in pairs or small parties with a lead male. Probably nomadic. Diet is vegetarian, feeding on grasses, shrubs, and *Acacia* spp. from which it can obtain most of its water needs. Potential predators largely exterminated in Egypt but hunting by humans would still be a major threat should the species still exist in Egypt. Has been known to associate with the Dorcas Gazelle. Gestation 156–169 days. Calf or calves (twins not uncommon elsewhere in range) born in winter or spring but no clear season as far as is known.

Similar species: See Dorcas Gazelle. The Mountain Gazelle has much shorter, stockier horns, is darker, and has distinct facial markings. The known ranges of the two species do not overlap.

MOUNTAIN GAZELLE (ARABIAN GAZELLE, CUVIER'S GAZELLE, EDMI)
Gazella gazella (Pallas, 1776)
Pl. 20
Subspecies occurring in Egypt: *G. g. arabica.*
Identification: Length 110cm–125cm; Tail 15–20cm; Shoulder 60–80cm; Weight 15–35kg. Rather large, slender gazelle with long limbs and short horns. Upper parts pale red-brown with distinct flank stripe of brown to gray-black. Color extends onto outside of

limbs. Underside and inside and back of limbs white. Rump white with border darker than rest of flank. Head conspicuously marked, similar to the markings of the Dorcas Gazelle. Ears rather short and grayer than body. Horns much shorter than previous species with the male's horns being a maximum 27cm, curved forward, with tips curved forward and

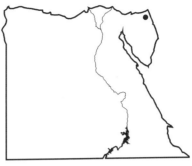

Mountain Gazelle
(*Gazella gazella*)

inward. Female's horns even shorter and more slender. Tail dark.

Range and status: Formerly over much of northwestern Africa and Arabia from the coast of Mauritania and Morocco to Algeria and Tunisia. Also Syria and Middle East to Arabian Peninsula. Everywhere decimated by hunting. In Egypt, known only from sight records from the 1930s in Wadi al-Arish. Status now uncertain. Listed as vulnerable by IUCN.

Habitat: Less strictly a desert gazelle than the Slender-horned Gazelle. Found in coastal desert, plains, savanna, and croplands. Also in mountainous areas in mountain wadis and vegetated slopes. Absent from true desert.

Habits: Unknown in Egypt but probably like other gazelles, active in morning and late afternoon and possibly at night. Lives in pairs or family parties led by an old male. Elsewhere, has been known to mix with Dorcas Gazelle. Diet consists of grass, leaves, shrubs, etc. May be able to obtain enough moisture from food to survive, but probably needs water. Potential predators of adults exterminated in Egyptian part of range but certainly vulnerable to hunting. Gestation about 6 months. Generally one calf.

Notes: As noted previously, gazelle classification is very complex and constantly changing. Some authors regard the population from northwestern Africa as a separate species, Cuvier's Gazelle or Edmi *G. cuvieri*. The current species remains the Mountain or Arabian Gazelle *G. gazella*. Harrison and Bates (1991) refer the Egyptian population—if one sighting warrants a population—to *G. g. cora*, but raise the question that records probably refer to the Dorcas Gazelle. **Similar species:** See other gazelles. Geographically separate from Slender-horned Gazelle.

IBEX *Capra ibex* Linnaeus, 1758
Pl. 21

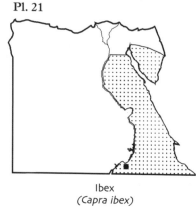

Ibex
(Capra ibex)

Subspecies occurring in Egypt: *C. i. nubiana* (Nubian Ibex). **Arabic:** *Taytal nubi* **Identification:** Length 155–195cm; Tail 15–25cm; Shoulder (male) 75–90cm, (female) 65–80cm; Weight (male) 60–80kg, (female) 50–70kg. Males significantly larger than females. Stout, heavyset goatlike animal with backswept horns that are particularly long and heavy in the male. Upper parts light brown above with rather grizzled appearance extending down upper half of each leg. Flanks similar color, underparts and inside of legs white. Distinctive leg pattern in both sexes with black patches above and below the knees, with the knees/pasterns themselves and the region above the hooves white. In males especially, black also along dorsal crest, chest, chin, and muzzle. Females less strongly marked. Males and some old females have a beard that is black. Ears similar in color to upper parts, inside black, with white margins. Horns present in both sexes but much more developed in the male. Long and backswept in a half or even three-quarter circle, they are flattened like sword blades and can reach 138cm in length along the curve. Thick and heavy, they have up to 30 prominent raised knobs on the front edge, less developed toward tip. Horns of female much smaller and smoother. From head on the horns sweep back and outward. Tail very short and black margined. Voice described as a 'whistling snort' when alarmed. Kids bleat.

Range and status: The Ibex ranges throughout the mountains of

Europe including the Pyrenees and Alps east to Russia, China and Mongolia, and south to India. Formerly, found over much of the Middle East including Syria, Israel, and Arabia. Also found in Sinai south through Egypt to Sudan, with an isolated population—now highly endangered—in Ethiopia. In Egypt, the Nubian Ibex was formerly widespread in Sinai and much of the Eastern Desert south to the Gebel Elba region. Today, it has been much reduced by hunting. Harrison and Bates (1991) estimated the total population in North Sinai to be 50 animals. It still occurs in declining numbers in the mountains of South Sinai where it is found in the St. Katherine Protectorate. Range in Eastern Desert now much fragmented and still victim of illegal hunting, preventing publication of exact locations of recent sightings. Largely restricted now to the remoter areas. Protected by Egyptian law but still hunted. As early as 1900, the decline in the Ibex population was noted and a sanctuary set up for them at Wadi Rishrash in the Eastern Desert. Also occurs in Gebel Elba Protected Area.

Habitat: The Ibex is almost entirely restricted to rocky, mountainous terrain, cliffs, canyons, and wadis. It is an extremely adept and agile climber.

Habits: Active mainly in early morning and late afternoon. Rest during heat of the day. Ibex, even close by, can be invisible against a rocky backdrop when motionless. Can wander over huge areas living in small herds of up to 40 or more animals, though generally in smaller groups of less than 10. Herds mixed male, female, and young, though females with young kids may separate. Group led by an old male. Also solitary. When moving, generally travel in single file. In Sinai especially, it can get very cold in the mountains. To preserve heat, they have been found following the sun around during their day's wanderings, starting on eastward-facing slopes. Conserve heat at night by sheltering amongst large boulders that have absorbed heat during the day. Diet is vegetarian, including grass, shrubs, *Acacia* spp., and roots. The Ibex needs standing water, a fact that hunters take advantage of at known watering sites. Sight, hearing, and smell all acute. Predators in Egypt largely exterminated. In Israel, the Ibex is the prime food of the Leopard. Calves vulnerable to a wider range of predators including birds of prey. Humans, however, are its greatest threat. In Israel, the rutting season is in October/November when the males fight over the females, the lead male mating most of the harem. Gestation 150–165 days with one kid being the norm though twins reportedly not uncommon.

Notes: Artifacts made from Ibex horn are still available in the Khan al-Khalili, the souk at Aswan, etc. They should not be bought as this only encourages the illegal hunting. Similarly, if the Ibex is to survive in Egypt, serious action should be taken against illegal trophy hunting.

Similar species: The Barbary Sheep *Ammotragus lervia* lacks the black and white markings on the legs, has a pronounced throat fringe, different coat color, and smooth, very differently shaped horns. The domestic goat is much smaller, longer haired, and very differently and variably patterned. Tracks, particularly in areas where the two coexist, can be hard to tell apart.

BARBARY SHEEP (AOUDAD, MANED SHEEP) *Ammotragus lervia* (Pallas, 1777)

Pl. 21

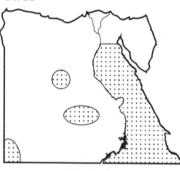

Barbary Sheep
(Ammotragus lervia)

Subspecies occurring in Egypt: *A. l. ornatus.*

Arabic: *Arwi, Kabsh gabali*

Identification: No measurements from Egypt. Length (male) 175–190cm, (female) 145–160cm; Tail (male) 20–25cm, (female) 15–20cm; Shoulder (male) 90–100cm, (female) 75–90cm; Weight (male) 100–140kg, (female) 40–55kg. Males considerably larger than females. An unmistakable, large, stocky, wild sheep with prominent mane along underside of neck and smooth, curved horns. Upper parts orange-brown to reddish, browner on legs. No flank stripe, underparts paler. Lower legs and feet pale, hooves dark. Profuse mane (more apparent in male than female) of long hair running from throat down underside of neck to the forelegs. Mane paler than upper parts. No beard as present in male Ibex. Head elongated, lacking distinctive pattern. Ears small and pointed, whitish inside. Horns large and heavy, often meeting at base on forehead. Curve out and back and then down and inward, almost forming a circle. Ringed, but lack the distinct knobs of the Ibex. Larger in the male, where they can reach 88cm along the curve, than in the female, where they rarely reach 40cm. Tail rather long-haired, even bushy. Naked on underside at base. Voice a series of grunts between female and calves and from rutting male.

Range and status: Formerly much of North Africa in mountainous areas from Mauritania to Morocco east through Algeria, Libya, and Egypt and south to Chad, Niger, and Sudan. Range now fragmented but still recorded in reasonable numbers in some regions. In Egypt, formerly over much of the Eastern Desert and the central and southern Western Desert including near Kharga, Farafra, and Gebel Uweinat. Now decimated by hunting and possibly extinct in the Western Desert (last record 1972). Anecdotal reports of the species from northeast of Wadi Allaqi in the south of the Eastern Desert where two were reportedly shot illegally by hunters in 1993. Protected by Egyptian Law, but this is poorly enforced. Listed under CITES Appendix II and categorized as vulnerable by the IUCN.

Habitat: Mountainous desert areas, cliffs, etc. Extremely adept climber. Descends to wadis to feed.

Habits: Little known in Egypt. As with most desert herbivores, active mainly in early morning and late afternoon/evening. By day seeks shade. Not known if the Barbary Sheep is territorial. Generally found in small family groups in the Eastern Desert of reportedly 5–6 animals. Family consists of 1 adult male plus females and calves and may amalgamate to form larger herds. Diet is vegetarian, feeding on grass, leaves, shrubs, and gourds. Probably less tied to water than Ibex and may be able to obtain enough moisture from food. Will drink if water is available. Predators largely exterminated but Leopard and Caracal *Felis caracal* may still be present in southern Eastern Desert. Humans have been greatest threat. Elsewhere, pairs in October/November. Gestation 150–165 days giving birth in spring to 1–3 lambs.

Notes: Ibex and Barbary Sheep probably used to extend over similar ranges in the past but the sheep could exploit areas with less available water.

Similar species: See Ibex, page 156.

The Rodents—Order Rodentia

With over 1,700 species currently recognized and new species still being described, the rodents make up the largest mammalian order. They occupy the entire range of habitats, from arid desert to dense rainforest, from artic snowfield to city center. Different species have become adapted to different lifestyles so that the rodents include the fossorial blesmols, the gliding anomalures, the aquatic beavers, the kangaroo-like Spring Hare *Pedetes capensis*, and a host of other forms. Yet, despite their omnipresence and apparent diversity, the rodents are a well-defined order. Their defining feature is their teeth, with two well-developed incisors in each jaw (the superficially similar lagomorphs have four), no canine teeth, and a gap between the incisors and the cheek teeth known as the diastema. The incisors grow throughout life and must be worn down through constant use. As far as basic body plan is concerned, one need look no further than the House Rat *Rattus rattus* for a typical rodent (there are over a thousand species of rats and mice alone in the family Muridae and few deviate far in appearance from the House Rat), though, as shown above, this basic design has proved highly adaptable.

Today, the rodents form one of the most successful mammal groups, not just in terms of number of species but in sheer numbers. They are opportunistic as well as adaptable and there are few corners of the globe where, courtesy of humans, such species as the House Mouse *Mus musculus* and the House Rat have not reached. This ability to invade all sorts of different habitats is matched by their rapid rate of reproduction. They generally have short gestation periods, large litters at short intervals, and the young are weaned rapidly and are capable of reproducing at an early age. In the House Mouse, for instance, gestation may be as short as nineteen days and up to fourteen litters a year have been recorded. This has several implications. They can rebound rapidly from catastrophic events, they can survive a high level of predation, a small number of pioneer individuals can rapidly colonize a new area, and they can adapt rapidly to changing environmental conditions. A further factor in their success has been their extremely wide-ranging diet.

Our own history has probably been more influenced by rodents than by any other group of mammals. Ratborne diseases have probably cost more—in terms of human life—than any war, uprising, or revolution. The Black Death, which decimated Europe in the fourteenth century, is just one example. Vast amounts of food are lost each year to rodent pests, either in the field or in storage. Yet, at the same time, two of their number, the hamster and the guinea pig, are today amongst the most popular pets.

In Egypt, the rodents are represented by thirty-three species in five families. For details, see the description that precedes each family. While the three alien commensals, the House Rat, the Brown Rat *Rattus norvegicus*, and the House Mouse, are all likely to be encountered around human habitation, cultivation, and detritus, the most widespread species are the gerbils and the jirds of the genera *Gerbillus* and *Meriones*, with ten and five representatives, respectively. They, together with a number of related but smaller genera, present the greatest identification challenge to the field naturalist in Egypt. Because the differences between them are generally slight, they have their own introductions.

The Mouse-like Rodents—Family Muridae (in part)

This large family contains some of the most widespread and familiar of the smaller mammals including such notorious commensals as the House *Rattus rattus* and Brown *Rattus norvegicus* Rats, and the House Mouse *Mus musculus*. Most members of the family share the broad features of this familiar trio: the large eyes and ears, long tails, and small size. However, the two features that distinguish this family from the other rodents are internal, namely, the structure of the molar teeth and of the chewing muscles.

Being such a large assemblage, they are inevitably divided into a series of fifteen more or less distinct subfamilies, though the number (and the number of species within each subfamily) is constantly being reviewed and changed. The following are represented in Egypt: the old world rats and mice—subfamily Murinae; and the gerbils—subfamily Gerbillinae.

The Gerbils—Subfamily Gerbillinae
81 species worldwide with 19 species in Egypt.

Along with the only distantly related jerboas, the gerbils form the most typical and numerous group of desert rodents in Egypt. Many of the species are very similar and identification of the gerbil, jird, and Fat Sand Rat *Psammomys obesus* species presents the biggest challenge to the mammologist in Egypt. Indeed, as with many rodent families, the number of species, even genera, recognized changes with each review.

The gerbils are all adapted to life in arid areas though they also occur in semi-desert, steppe, salt marsh, and agricultural regions. Habitat can be important for identification as it will be seen that certain species seem to occur with each other and are associated with a specific habitat. Zoologists, in trying to understand the distribution of the gerbils, have recognized three broad groups: those of the genus *Tatera* that are found in the savanna and steppe regions of sub-saharan Africa; the *Gerbillus*-type gerbils of the true deserts of the Horn of Africa, North Africa, and southwestern Asia; and the *Meriones*-type gerbils of Central Asia where the temperature can drop below freezing. The last two groups overlap in the Middle East and both are well-represented in Egypt.

All gerbils show adaptations to life in the desert. Their main problem lies in combining small size with life in an arid environment. Having a large surface area to volume ratio, they are in danger of losing water very rapidly. The gerbils overcome this problem through a series of behavioral and metabolic adaptations. Virtually all gerbils, with the notable exception of the Fat Sand Rat, are predominantly nocturnal. They spend the day in burrows of varying complexity that may or may not be plugged. The burrow is considerably cooler than the surface and the temperature relatively stable throughout the day and night. The gerbils emerge at night to feed. All gerbils are vegetarian, eating grasses, shoots, and seeds. As gerbils often live in areas devoid of fresh water, they have to obtain all the moisture they require from their food. Even dry seeds, foraged at night, may be covered in dew, and thus, by being nocturnal, the gerbils can increase the water content of their food. Their burrows are often much more humid than the surface and this further enhances the moisture content.

The gerbils not only maximize the amount of moisture they obtain from their diet, they also minimize the amount of moisture they loose. Their urine is extremely concentrated due to highly efficient kidneys, and their feces are very dry.

In adapting to a nocturnal mode of existence, the gerbils have developed a number of common features. All are mouse-like in form with well-developed hind legs and forelegs that can be used as hands. The tail is longish in most species (though short and stubby in the Fat-tailed Jird *Pachyuromys duprasi)* and often has a terminal tuft or dark band. Also, there is often a white patch on the rump and behind the ears that may be useful in recognition at night. The eyes are very large, the ears quite large and rounded, and the vibrissae well-developed. The coat is very soft and dense, frequently with a more or less defined flank stripe. The color is variable, even within species, and often conforms to the color of the habitat. The soles of the feet may be naked as in many species of stony desert and steppe, or haired as in species of sandy areas.

Identifying the Egyptian gerbils can be very difficult, even in the hand, and very few species are distinctive enough to be identified by a brief sighting. Their nocturnal habits and frequently remote haunts further complicate matters. For the casual observer, range and habitat type can be enormously important in working out what sort of a gerbil has been seen or what species is likely to have left the myriad of little tracks around the campsite over night. Also, some burrows can be identified and one can refer to likely associated species. In the hand, tail length with absence, presence, color, and extent of a terminal tuft is important, as is the color of the flank stripe. The prominence and extent of white rump and ear patches, ears, feet and claw color, and color of vibrissae can also be important as well, of course, as overall size and color.

Gerbil remains are frequently found in the pellets of desert birds of prey, perhaps most notably Eagle Owls *Bubo bubo*. While it is beyond the scope of this guide to provide details on cranial characteristics, both Osborn and Helmy (1980) and Harrison and Bates (1991) provide excellent descriptions and identification tables.

GIZA GERBIL (CHARMING DIPODIL) *Gerbillus amoenus* (de Winton, 1902)
Pl. 23

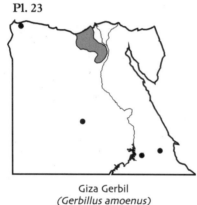

Giza Gerbil
(Gerbillus amoenus)

Subspecies occurring in Egypt
G. a. amoenus.
Arabic: *Yarabil aminas*
Identification: Length 163–216mm; Tail 93–116mm Weight 10.7–17.5g. A small gerbil with a long tail (one and one-third body length) Above, dark yellow-brown more yellow on side, especially in darker individuals. Fur short. Below, including underside of tail and limbs, white Distinctively large, whitish rump patch. Facial pattern with darkish streak running from below the eye to the base of the ear. White patch above each eye and behind each ear distinct. Tail long, white below and brownish above with indistinct tuft on last quarter to third. Soles of feet pale and hairless. Feet white above.

Range and status: North Africa from Mauritania and Morocco east, but doubtful in Tunisia. In Egypt, found west of the Delta including Wadi Natrun and the Fayoum, where reportedly common Also recorded from southern Eastern Desert including Wadi Allaq and Dakhla in the Western Desert. Disjunctive distribution would imply it is more widespread than records suggest.

Habitat: Reported from a wide range of habitats including salt marshes, canal banks, cultivated areas, wasteland, sandy and rocky (*hamada*) desert, tamarisk stands, and palm groves in oases. Also recorded from settlements in the Fayoum and tents in desert areas.

Habits: Nocturnal, spending the day in a shallow, branchless burrow, dug in a variety of substrates, from loose sand to hard soil. In palm groves, in sand beneath dead palm fronds. Specific diet not known in Egypt but probably similar to other gerbil species, i.e., seeds, etc.

Associated species: In salt marshes with the Fat Sand Rat (which is diurnal), Pygmy Gerbil *Gerbillus henleyi*, Large North African Gerbil *Gerbillus campestris*, Lesser Short-tailed Gerbil *Dipodillus simoni*, Four-toed Jerboa *Allactaga tetradactyla,* and Greater Egyptian Jerboa *Jaculus orientalis*. In sandy areas with the Lesser Egyptian Gerbil *Gerbillus gerbillus* or Anderson's Gerbil *Gerbillus andersoni*.

Similar species: Probably not distinguishable in the field from Wagner's Gerbil *Gerbillus dasyurus* or Baluchistan Gerbil *Gerbillus nanus*. Both of these also occur in the Sinai where the Giza Gerbil is absent. Distinguished from the jirds by proportionately longer tail with more of a terminal brush, more distinct white rump patch, and clearer white patch above the eye.

ANDERSON'S GERBIL *Gerbillus andersoni* de Winton, 1902
Pl. 24

Subspecies occurring in Egypt: *G. a. andersoni, G. a. inflatus,* and *G. a. bonhotei.*

Arabic: *Garbil andarsun*

Identification: Length 193–265mm; Tail 110–150mm; Weight 15.9–38.4g. A medium-sized gerbil of rather slender build. Coat long and dense, orange-brown above, clear orange along the flanks, and clearly demarcated white below, including underside of

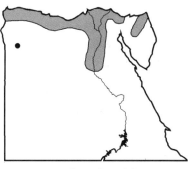

Anderson's Gerbil
(Gerbillus andersoni)

tail, except at base, which is buffish. Ears rather large and pigmented at tip, white patch at back of ear less distinct than in other *Gerbillus* spp. Eyes have a thin, black, orbital ring, but white patch above eye indistinct. Broad band of dark hairs runs from below each eye to the base of the ear. In the hand, the soles of the feet are haired except for a small naked patch in the hind soles. Soles pigmented. Tail long, colored as upper side of body and white below sharply demarcated. White rump patch small. Tuft on last quarter of the tail is indistinct and brownish.

Range and status: North Africa from Libya east to Sinai, Israel and Palestine, and northern Arabia. In Egypt, restricted to the northern portion of the country. *G. a. andersoni* found from the Libyan border along the Mediterranean coastal desert to Alexandria. Also Siwa and Wadi Natrun. *G. a. inflatus* found along the Delta's northern and western margins south to environs of Cairo, Saqqara, and the Fayoum. *G. a. bonhotei* recorded only from northeastern Sinai.

Habitat: Essentially a gerbil of sandy areas though has been recorded elsewhere. Along Delta, Nile Valley, and the Fayoum found in sandy stretches, cultivated and uncultivated, and sandy areas between palm groves. Along Mediterranean, found in vegetated sandy dune areas and sandy margins to salt marshes.

Habits: Nocturnal, spending day in burrow often dug at the base of a bush. Diet in Egypt uncertain. Elsewhere seasonal, in winter feeding largely on greenery, switching to seeds in spring. Also known to take insects. Breeding season probably extended with records from September to June. Litter size 3–7.

Associated Species: Anderson's Gerbil has been found in association with the Lesser Egyptian Gerbil, Pallid Gerbil *Gerbillus perpallidus*, Shaw's Jird *Meriones shawi*, Giza Gerbil, and Lesser Short-tailed Gerbil. Also reported with Large North African Gerbil and House Mouse, and possibly Greater Egyptian Gerbil *Gerbillus pyramidum*, Greater Egyptian Jerboa, and Fat Sand Rat.

Notes: In Israel, Anderson's Gerbil has been found to compete successfully with the much larger Tristram's Jird *Meriones tristrami*, and the two are rarely found in the same area. *G. a. bonhotei* is sometimes considered a full species.

Similar species: Other gerbils but especially the Lesser Egyptian Gerbil, with which it may associate. The Lesser Egyptian Gerbil is superficially very similar. However, it is paler with unpigmented ears and soles, a much larger white rump patch, a distinct white patch behind ear, and tail with large pale brush.

LARGE NORTH AFRICAN GERBIL (LARGE NORTH AFRICAN DIPODIL)
Gerbillus campestris (Levaillant, 1857)
Pl. 24
Subspecies occurring in Egypt:
G. c. wassifi, G. c. haymani, G. c. patrizii, and *G. c.venustus.*
Arabic: *Yarabil shamal Ifriqiya al-kabir*
Identification: Length 204–278mm; Tail 118–153mm; Weight 21.3–44.1g. A rather variable gerbil with prominent ears. Above, orange to orange-brown that can appear streaked in paler individuals. Pales toward flanks to a nar-

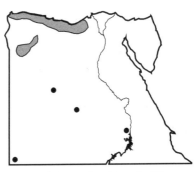

Large North African Gerbil
(Gerbillus campestris)

row orange border that extends onto the limbs. White below. Head with large ears that are completely pigmented. White patch behind ear and above eye, indistinct hairs that are dark tipped. White rump patch indistinct and may be absent. Tail long, orange to orange-brown above, variable below from white to brownish, thus, may be uniform or bicolored. Brush one-half to one-third tail length but indistinct. In hand, soles of feet are naked.
Range and status: North Africa from Chad and Niger north to Morocco and east to Egypt and Sudan. In Egypt, *G. c. wassifi* found along the Mediterranean coastal desert from Sallum to Alexandria, *G. c. haymani* in the Western Desert including Siwa, Farafra, and the Qattara Depression, *G. c. patrizii* restricted in Egypt to Gebel Uweinat, and *G. c. venustus* only from the Nile Valley southwest of Aswan. *G. c. venustus* might be extinct in Egypt since the area in which it was found is now inundated by Lake Nasser.
Habitat: Has a wide habitat tolerance. Recorded from sea cliffs along the north coast, quarries, old buildings, temples, acacia groves, sand dunes, rocky gullies, palm and fig groves, barley fields, salt pans, lake edges, sedge thickets, cultivated areas, etc. In Libya, regarded as the most widespread rodent species.
Habits: Nocturnal. Habits little known in Egypt. Burrows tend to

be in sandy soil but also in and amongst rock crevices. Breeding unclear and probably varies with populations. Along the north coast, breeding reported during the winter rainy period, but elsewhere, litters have been found in late spring. In northern Sudan, the species is reported to breed from September to November. Litter size probably 3–6.

Associated Species: Because of its wide habitat tolerance, the Large North African Gerbil has been found with a variety of species. In sandy areas, it has been recorded with Anderson's Gerbil and Lesser Egyptian Gerbil. In rocky areas, it has been found with the Cairo Spiny Mouse *Acomys cahirinus* and Middle Eastern Dormouse *Eliomys melanurus*, and in salt marshes, with the Fat Sand Rat, Giza Gerbil, Lesser Short-tailed Gerbil, and Pygmy Gerbil.

Similar species: Other gerbils. The Lesser Egyptian Gerbil is smaller with distinct white rump patch and clearer white markings behind ears and above eyes.

WAGNER'S GERBIL (WAGNER'S DIPODIL, ROUGH-TAILED DIPODIL, WADI HOF GERBIL) *Gerbillus dasyurus* (Wagner, 1842)
Pl. 25

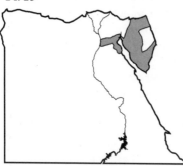
Wagner's Gerbil
(*Gerbillus dasyurus*)

Subspecies occurring in Egypt: *G. d. dasyurus*.

Identification: Length 189–248mm; Tail 109–136mm; Weight 16–34.9g. A medium-sized, rather slender gerbil. Coat long and silky. Above, rather pale yellow-brown without orange tinge, and very finely speckled black due to black tips to hair. Below white. Forelimbs, upper lip, and inner hind limbs white. White rump patch absent. Ears smaller than in previous species, pigmented, and fawn-gray. White patch at back of ear and above eye variably distinct. Narrow, black, orbital ring around eye. Tail long, yellow-brown above, white below with distinct tuft that extends along the terminal half of the tail. In the hand, soles of feet are naked.

Range and status: Arabia, the Middle East north to Syria and west to Egypt. In Egypt, recorded throughout Sinai north of Wadi Kid, as well as the northern region of the Eastern Desert south to Wadi Iseili south of Ain Sukhna, and west to Muqattam and Wadi Hof.

Habitat: Primarily a gerbil of rocky areas, cliffs, quarries, scree and boulder slopes, and rock-strewn wadi floors. In South Sinai, where it is not found below 500m, burrows found in sandy areas in rocky wadis, but also in the garden of St. Katherine's Monastery. Also recorded from margins of cultivated areas and salt flats.

Habits: Strictly nocturnal. Spends day in burrow that may be a shallow hole beneath a rocky overhang, or a long and very complex burrow in sandy areas between rocks. Burrow sealed with sand during the day to maintain cooler temperature. Feeds on desert herbs and grasses, though insects also recorded. Elsewhere has been recorded as feeding on snails, and burrows may be identified by the accumulations of snail shells outside them. Essentially solitary and may wander long distances from the burrow in order to find food. Breeding has been recorded more or less throughout the year with the female giving birth to 2–6 young.

Associated Species: Wagner's Gerbil has been found with the Bushy-tailed Jird *Sekeetamys calurus*, Cairo Spiny Mouse, Golden Spiny Mouse *Acomys russatus*, and Middle Eastern Dormouse. In Yemen, it has been found to share burrows with the Libyan Jird *Meriones libycus* and in northeastern Saudi Arabia, both Wagner's Gerbil and the Baluchistan Gerbil have been found in association with the Libyan Jird.

Similar species: Other gerbils. Wagner's Gerbil differs from the Large North African Gerbil and Anderson's Gerbil by the absence of orange in the coat and the much longer and more prominent tail tuft. From the Giza Gerbil, it can be told by the absence of a white rump patch (large and distinct in the Giza Gerbil) and the fully pigmented ears.

LESSER EGYPTIAN GERBIL (LESSER GERBIL) *Gerbillus gerbillus*
(Olivier, 1801)
Pl. 23

Lesser Egyptian Gerbil
(Gerbillus gerbillus)

Subspecies occurring in Egypt:
G. g. gerbillus, G. g. asyutensis,
and *G. g. sudanensis.*
Arabic: *Bayyud*
Identification: Length 167–
247mm; Tail 91–137mm;
Weight 13.6–34.7g. A small,
slender-limbed gerbil, generally pale though variable.
Above, pale orange-yellow to
reddish orange, paling along
the sides, though color not extending to limbs except to the
thigh. Demarcation distinct. Forelimbs, underparts, and much of face,
white. White rump patch large and distinct. Tail long and bicolored
along entire length, same color as upper parts above, white below.
Terminal brush about one-third total tail length, moderately developed but thinly haired. Hind feet elongated and, in hand, soles almost
entirely haired and unpigmented. Ears unpigmented, though margin
blackish, and proportionately rather small, length being less than half
that of the hind foot. White patch behind ear distinct. Whiskers long,
mixed black-and-white.
Range and status: A North African species found from Mauritania
east to Egypt and Sudan, and south to Uganda. Also Sinai, parts of
Israel, and Jordan. In Egypt, one of the most widespread desert
rodents found virtually throughout the country, including Cairo and
its environs. *G. g. gerbillus* is the subspecies occurring throughout the
Western Desert to the west bank of the Nile and the western margin
of the Nile Delta. It has been recorded from all the major Western
Desert oases including Wadi Natrun and the Fayoum, where it is
reportedly common. *G. g. asyutensis* is found in Sinai and the Eastern
Desert west to the east bank of the Nile, and the eastern margin of
the Delta. It occurs south, to a line running roughly from Edfu to
Marsa Alam. South of Marsa Alam, it is replaced by *G. g. sudanensis.*

Habitat: A gerbil of sandy areas ranging from duneland, including coastal dunes and salt pans, to sandy wadi beds. Also sandy areas in cultivated land and sandy tracts amongst palm groves, reeds, etc. Attracted to campsites. Not reported above 1,100m in mountainous areas.

Habits: Nocturnal. Spends day in burrow that is sealed with a sand plug especially during the hotter part of the year. Burrow in desert areas in flat tract of sand and, even in semi-desert, generally clear of vegetation, and up to 80cm deep. Diet includes seeds, leaves, grasses, etc. Attracted to camel dung that it breaks up to search for undigested food items. Breeding probably between January and May with 3–6 young in a litter.

Associated Species: Found in the same type of terrain as Anderson's Gerbil, Greater Egyptian Gerbil, and Pallid Gerbil. In Israel, a population was studied on sand dunes that adjoin saline pans. This particular population co-existed with the Baluchistan Gerbil, though the latter was rarely recorded away from the salt flats and the former was never found away from the dunes. In the Eastern Desert, it has been found on boulder-strewn, sandy wadi floors with the Cairo Spiny Mouse.

Notes: One of the most widespread Egyptian mammals.

Similar species: Other gerbils. See also Anderson's Gerbil and Large North African Gerbil. Smaller than the Pallid Gerbil.

PYGMY GERBIL (HENLEY'S GERBIL, PYGMY DIPODIL) *Gerbillus henleyi* (de Winton, 1903)

Pl. 23

Subspecies occurring in Egypt: *G. h. henleyi* and *G. h. mariae*.

Identification: Length 134–174mm; Tail 72–99mm; Weight 7.2–11.4g. Tiny gerbil with dense, silky fur. Size alone should distinguish it. Buff-brown above, darker along the back, paler on the sides. Forelimbs and underside white. Feet white, soles hairless and unpigmented. Tail long, buff-brown above and white below. Tail brush indistinct and a quarter or less of the total tail length. Limbs long and slender, hind feet elongated. Ears with white patch at base of back. Moderately-sized and unpigmented. Clear white patch above the eye. White rump patch prominent.

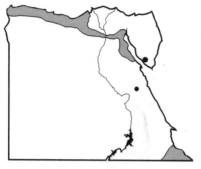

Pygmy Gerbil
(Gerbillus henleyi)

Range and status: A predominantly North African species found from Algeria east to Egypt and Sudan, Sinai, and east to Israel, Jordan, and western Arabia, with one isolated record from Oman. In Egypt, *G. h. henleyi* is found in the Mediterranean coastal desert from Sallum east to Alexandria and along the western margin of the Delta, including Wadi Natrun, south to Cairo. *G. h. mariae* is found patchily through the Eastern Desert, with most records coming from the northernmost regions, including Wadi Digla, and southernmost regions as well as in Sinai, including Wadi Feiran, Gebel Maghara, and Ras al-Naqb near Taba.

Habitat: Varied. Adapted for life in very arid areas but most often recorded from areas with moderate amounts of vegetation. This may be a consequence of where most research has been done. Found in vegetated rocky wadi beds, vegetated plains, grassland, coastal marshes, salt marshes, cultivated land, and reclaimed farmland.

Habits: Nocturnal. Burrow is shallow, simple, and excavated in sand. The entrance is characteristically small, likened by one author to that of a lizard, being only 1–2cm in diameter. Burrow entrances are plugged during the day. Diet probably seeds and other vegetation. Breeding poorly known, but litters have been found in June and August in Egypt. Litter size around 4.

Associated Species: The Pygmy Gerbil has been found in the same habitat as the Lesser Short-tailed Gerbil, Giza Gerbil, Lesser Egyptian Gerbil, and Greater Egyptian Gerbil. It is known to use the old burrows of the Fat Sand Rat in the Western Desert, and elsewhere has also been trapped in front of the burrows of the Silky Jird *Meriones crassus* and the Libyan Jird.

Similar species: Other gerbils are all larger than the Pygmy Gerbil.

MACKILLIGIN'S GERBIL (MACKILLIGIN'S DIPODIL) *Gerbillus mackilligini* Thomas, 1904
Pl. 25

Subspecies occurring in Egypt: *G. m. mackilligini.*

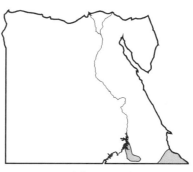

Identification: Length 171–224mm; Tail 99–138mm. Small, little-known gerbil with a long tail and prominent ears. Yellow-brown above becoming more yellow along the sides extending to fore- and hind limbs. White below. Feet white. Soles naked and unpigmented. Ears largish and pigmented. Dark band runs from below the

Mackilligin's Gerbil
(Gerbillus mackilligini)

eye to the base of the ear. White patches above eye and behind ear indistinct. White rump patch absent. Tail dark above (darker than body) and paler below. Terminal brush distinct and up to half the tail length.

Range and status: North Africa. In Egypt, only recorded from the southern Eastern Desert with records from Gebel Elba, Wadi Allaqi, and just south of Aswan.

Habitat: All Egyptian specimens have been obtained from vegetated, particularly grassy areas, often in the vicinity of water, or from around ruins.

Habits: Nothing known in Egypt other than that it plugs its burrow during the day. Presumably nocturnal.

Associated Species: Not known.

Notes: The classification of Mackilligin's Gerbil has been the subject of some debate. It has been previously classified as a subspecies of both the Baluchistan Gerbil and Wagner's Gerbil and has only been widely recognized as a separate species since 1959.

Similar species: Very difficult to distinguish from the Baluchistan Gerbil and Wagner's Gerbil. From the former, note the pigmented ears, absence of white rump patch, and very long tail brush. From the latter, only safely identified on skull characteristics, but the two species are separated geographically.

BALUCHISTAN GERBIL *Gerbillus nanus* Blandord, 1875
Pl. 25

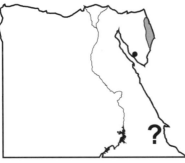

Baluchistan Gerbil
(Gerbillus nanus)

Subspecies occurring in Egypt: uncertain but probably *G. n. arabium.*

Identification: Length 140–235mm; Tail 80–145mm. A slender, medium-sized gerbil that closely resembles Wagner's Gerbil. Yellow-brown above, tinged gray due to gray bases to hairs. Flanks less gray. Under-parts, forelegs, inner hind legs, and feet are white. Soles of feet unpigmented and virtually naked. Coloration sharply demarcated. Ears medium-sized, with only the tip pigmented. White patches above eye and behind ear distinct. White rump patch large and distinct. Tail is long, yellow-brown at base becoming grayer toward the tip. Below whitish. Terminal tuft small but distinct, rather variable in color but normally grayish.

Range and status: Algeria east across North Africa to Somalia, Sudan, and Egypt. Sinai east throughout the Arabian Peninsula to Iraq, Iran, and western Pakistan. In Egypt, recorded in Sinai and the southeastern Eastern Desert. Status uncertain—see notes.

Habitat: Uncertain in Egypt. Elsewhere it occurs in a wide range of habitats from salt flats and well-vegetated, semi-desert, to sandy desert. Recorded at altitudes from sea level to 1,000m.

Habits: Unknown in Egypt. Elsewhere, active at dusk and at night using regular trails that take full advantage of any available cover. May wander long distances in search of food. Breeding probably winter and early spring.

Associated Species: Unknown.

Notes: The status of the Baluchistan Gerbil in Egypt is uncertain due to identification problems. Specimens taken by Wassif and Hoogstraal (1954) in South Sinai were assigned to *Gerbillus quadri maculatus*, a name now regarded as obsolete. The specimens were reidentified as *G. nanus*. However, further study assigned them to Wagner's Gerbil, a classification now generally accepted. The species was not included in Osborn and Helmy (1980), but according to Harrison and Bates (1991) there is at least one specimen from South Sinai in the British Museum that can be assigned to the Baluchistan Gerbil. The origins of the Gebel Elba records are uncertain, but as the species ranges south to northern Somalia in northeastern Africa, they are not unexpected.

Similar species: Other gerbils. Very similar to Wagner's Gerbil, but note large white rump patch absent in that species. Also tail brush, while distinct, is much shorter (up to half the length of the tail in Wagner's Gerbil).

GREATER EGYPTIAN GERBIL (GREATER GERBIL) *Gerbillus pyramidum*
I. Geoffroy St-Hilaire, 1825
Pl. 23

Subspecies occurring in Egypt: *G. p. pyramidum, G. p. floweri, G. p. geddedus,* and *G. p. elbaensis.*

Arabic: *Damsi*

Identification: Length 230–393mm; Tail 128–180mm; Weight 14–18g. Large gerbil, stout bodied but long limbed. Rather variable. Above, orange-brown to brown, darker along back, paler along the flanks and at base of tail

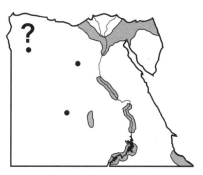

Greater Egyptian Gerbil
(Gerbillus pyramidum)

where hairs have white roots. Underparts, inner hind legs, forelegs, and feet white. Tail long, similar in color to upper parts, though sometimes browner, underside white or whitish. Tail bush varies from black to brownish and is around one-third the total length of the tail. Eyes large. Pale patch above eye and behind ear variable, sometimes indistinct. White rump patch may be absent. Soles of feet haired, except for small patch on rear soles.

Range and status: North Africa from Morocco south to Senegal

east to Egypt, Sudan, and Somalia. Also Sinai east to Israel and
Palestine. In Egypt, widespread. *G.p. pyramidum* is found in the
Western Desert to the western margin of the Delta, including Wadi
Natrun, south along the Nile Valley to the Sudanese border on both
banks of Lake Nasser. It has been found around the Giza pyramids.
Recorded in Siwa though record is questionable, and from the
Fayoum where it is reported to be common. *G. p. gedeedus* has been
recorded from the Western Desert oases (other than Siwa). *G. p.
elbaensis* has only been recorded from the southeasternmost corner
of the Eastern Desert, especially the Gebel Elba region. *G. p. floweri*
has been recorded from North Sinai and the northern Eastern
Desert (not recorded south of Suez or in Sinai south of Wad
Wardan near Ras Sudr). Does occur in the dunes of South Sinai
where *G. gerbillus* is common.

Habitat: Varied but not a gerbil of open deserts. In the Western
Desert, for example, the distribution is restricted to oases. Found in
palm groves, cultivated areas, vegetated wadi floors, sandy tracts
near cultivated areas, in farmland, and tamarisk scrub at desert margin in the Fayoum. In more barren areas, found around buildings. In
oases, around deserted buildings, cultivated areas, cisterns, palm
groves, etc.

Habits: Nocturnal, in Sudan coming out almost at sunset, and
reportedly high-strung and nervous, therefore, even more difficult
than other species to see. Burrows can be extensive and are plugged
during the day. Diet includes seeds and other vegetation. Will tear up
camel droppings for digestible items. Camel droppings, and other
food items, have been found stored in burrows. When threatened
runs for the burrow or seeks thick cover. Breeding in Egypt uncertain. In Sudan, occurs from June through to March. Gestation 22
days. Litter size probably 3–5.

Similar species: Other gerbils. Distinguished from Lesser Egyptian
Gerbil by larger size, lack of (or much smaller) white rump patch,
longer tail with more distinct tuft, and overall darker color.
Distinguished from Anderson's Gerbil by larger size and tail characteristics, and from the Pallid Gerbil, by much darker coloring and
smaller, or absent, white rump patch.

PALLID GERBIL *Gerbillus perpallidus* Setzer, 1958
Pl. 24

Subspecies occurring in Egypt:
G. p. perpallidus.
Identification: Length 223–
267mm; Tail 128–150mm;
Weight 26.2–48.4g. Medium-
sized, pale-colored gerbil.
Pale orange above, uniform
with no darker stripe down
the back, though hairs at base
of tail dark-tipped. Underside, forelimbs, and feet
white. White patches behind
ear and above eye distinct.

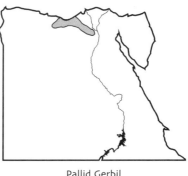

Pallid Gerbil
(Gerbillus perpallidus)

White rump patch distinct. Tail long, pale orange above, white
below with indistinct tail brush on terminal third. Ears unpigmented.
Soles haired.

Range and status: Endemic to Egypt. Recorded from the northeastern Western Desert, including the northeasternmost corner of the
Qattara Depression, east along the Mediterranean coastal desert
almost to Damietta. Also south along the western margin of the
Delta, including Wadi Natrun, and south to Abu Rawash. A little-
studied species and status unknown.

Habitat: Generally sandy areas from coastal dunes to sandy margins,
to agricultural areas, and sandy tracts in acacia groves, etc. In Wadi
Natrun, recorded from sandy and muddy lake shores with vegetation.

Habits: Poorly known. Presumably nocturnal. Burrows in sand.
Breeding probably in April/May. Litter size unrecorded.

Associated Species: Has been recorded in the same areas as the
Lesser Egyptian Gerbil, Anderson's Gerbil, Greater Egyptian Gerbil,
Lesser Egyptian Jerboa *Jaculus jaculus*, Libyan Jird, and Shaw's Jird.

Similar species: Differs from Lesser Egyptian Gerbil in larger size,
less conspicuous tail brush, and lack of black margin to ear.
Distinguished from Anderson's Gerbil by larger size and paler coloring, especially down the center of the back. Can be very similar to the
Greater Egyptian Gerbil. From *G. p. pyramidum* (the subspecies of

the Greater Egyptian Gerbil whose range overlaps that of the Pallid Gerbil) told by paler color above, large white rump patch, lack of dark stripe down the center of the back, and paler tail tuft. From *G. p. floweri*, virtually indistinguishable except on skull characteristics. However, this subspecies of the Greater Egyptian Gerbil is geographically separated from the Pallid Gerbil only occurring in the northern Eastern Desert and North Sinai.

LESSER SHORT-TAILED GERBIL (SIMON'S DIPODIL) *Dipodillus simoni* Lataste, 1881
Pl. 24

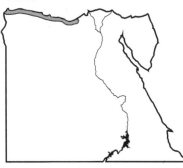

Lesser Short-tailed Gerbil
(*Dipodillus simoni*)

Subspecies occurring in Egypt: *G. s. kaiseri*.

Identification: Length 163–216mm; Tail 93–116mm; Weight 10.7–17.5g. A small gerbil, total length not much greater than the Pygmy Gerbil but with a proportionately shorter tail. Above, pale yellow-brown. Flanks paler with yellow strip along the side, not extending onto forelegs. Underparts and inner limbs white. Feet white above and below. Soles naked and hind feet unpigmented. Head colored as upper parts but with distinct broad, darker band through eye to base of ear. Whitish patch above eye present but not conspicuous. White patch behind the ear large. Ear pigmented. Lacks white rump patch. Tail proportionately short compared with other gerbils and jirds, being barely longer than the head and body length. Buffish with some black hairs on upper surface. Brush barely discernable.

Range and status: North Africa from Algeria east to Egypt. In Egypt, range restricted only to the north coast, with records from Mersa Matruh to the western margin of the Nile Delta near Alexandria.

Habitat: In Egypt, found in and around salt marshes. Also in olive groves and other cultivated areas. Elsewhere, also recorded from mountain plateaus and vegetated desert.

Habits: Little known. Nocturnal. The burrows recorded from salt marshes were unplugged. Diet not known. Breeding recorded in October in Tunisia. Litter size around 4.

Associated Species: In salt marshes, occurs with the Giza Gerbil, Pygmy Gerbil, Large North African Gerbil, Fat Sand Rat, Four-toed Jerboa, and Greater Egyptian Jerboa.

Notes: The taxonomy of the Lesser Short-tailed Gerbil is unclear; some authors split the species into [*Dipodillus*] *simoni* and [*Dipodillus*] *kaiseri*. In view of the overlap of features between the two, this field guide follows Osborn and Helmy (1980) in treating the two as one species [*Dipodillus*] *simoni* with the subspecies *kaiseri* found in Egypt.

Similar species: Other gerbils, but differs from all by the proportionately shorter tail which lacks a terminal brush. Of those gerbil species occurring in the same habitat, it further differs from the Giza Gerbil and Pygmy Gerbil because it lacks a white rump patch. Also distinguished from the Giza Gerbil by less distinct head markings and pigmented ear (not just tip). From Pygmy Gerbil, told by larger body size, yellowish (not buff) upper parts, and pigmented ear. From the Large North African Gerbil, told by much smaller size and yellow (rather than orange-brown) upper parts.

SILKY JIRD (LITTLE SAND JIRD, SUNDEVALL'S JIRD) *Meriones crassus* Sundevall, 1843

Pl. 26

Subspecies occurring in Egypt: *M. c. craasus*, *M. c. pallidus*, and *M. c. perpallidus*.

Arabic: *Maryunaz krasus*

Identification: Length 219–311mm; Tail 105–158mm; Weight 51–112.8g. Large but lightly-built jird. Above, pale yellow-brown, though variable, indistinctly speckled black with a gloss apparent in good light along the back. Fur long and silky. Below, inside limbs and feet white. Demarcation distinct. Top of head dark, paler buff on cheeks and pale around eye with white patch above. Distinct white patch behind ears. Ears moderately-sized and sparsely coated with white hairs. Tail long, yellow-brown above and white to pale below. Tuft distinct, black and covering terminal third of tail.

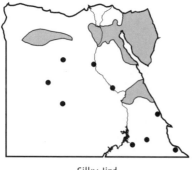

Silky Jird
(Meriones crassus)

Underside of tip white. Forefeet naked. Hind feet haired apart from naked patch on sole. Claws pale.

Range and status: North Africa from Morocco to Egypt, south to Sudan and northern Nigeria. Sinai and much of the Middle East. East to Iraq, Iran, Afghanistan, and western Pakistan, and north to eastern Central Asia. In Egypt, widespread. *M. c. crassus* found in Sinai, where absent from the coastal desert and above 1,500m (recorded from St. Katherine's Monastery), and the Eastern Desert from the southeastern margin of the Delta south to Marsa Alam. One isolated record from the northern border of the Sudanese Government Administration Area at Shalatin. *M. c. perpallidus* is found in the Western Desert, including the major oases, the margins of the Qattara Depression, and Siwa. Occurs south to Kharga and north to the western margin of the Delta, including Wadi Natrun. Reportedly common in the Fayoum. Absent from the Mediterranean coastal desert. *M. c. pallidus* recorded from the southern Eastern Desert south of Aswan.

Habitat: Varied but generally near vegetation or settlements. Has been recorded in rocky wadi beds, from palm groves, encampments, acacia groves, and rushes. In Sinai, found up to 1,500m in St. Katherine. Also recorded in colonies around rubbish tips.

Habits: Largely nocturnal but also active at dusk and even during the day. Lives in colonies in complex burrow systems, often burrowed into very hard substrate. Burrows normally shallow with a number of openings and not plugged during the day. Burrows may be around clumps of vegetation, e.g., Donkey Melons *Citrillus colocynthis*. Diet is broad, consisting of seeds, green vegetation, leaves, fruits, and berries. Also reported to eat locusts and scraps from Bedouin camps. In Egypt, breeding season probably November to June. Gestation 22–24 days and litter size 1–6.

Associated Species: The Silky Jird has been found in the same habitat as the Lesser Egyptian Jerboa, the Lesser Egyptian Gerbil, and the Pallid Gerbil. Also with Shaw's Jird and Libyan Jird.

Notes: Some authors recognize a further subspecies from Egypt *M. c. asyutensis.*

Similar species: Other jirds but paler and the only jird with a large white patch behind the ear.

LIBYAN JIRD *Meriones libycus* (Lichtenstein, 1823)
Pl. 26

Subspecies occurring in Egypt: *M. l. libycus.*

Identification: Length 238–312mm; Tail 115–157mm; Weight 65–108.8g. Large jird with distinct black tail tuft. Yellow to orange-brown above, dark with dense, soft fur. Underparts white with a narrow orange line running along the demarcation and onto wrist and thigh. Feet white. Soles of hind feet par-

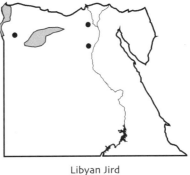

Libyan Jird
(Meriones libycus)

tially haired. Claws black. Head largely grayish with white around eye. Ears moderately sized. Unpigmented and with small, whitish patch behind base of each ear. Tail long, yellow to orange-brown above, not significantly bicolored, and dark orange at base. Tail brush distinct, black, reaching up to half the length of the tail.

Range and status: From Libya east to northern Egypt, Sinai, northern and eastern Arabia, the Middle East, north to Syria, east to Iraq and Iran. In Egypt, largely restricted to the northern Western Desert and margins of the Qattara Depression. Only reaches the Mediterranean coastal desert at Sallum. Also, Wadi Natrun and a few records from Wadi al-Rayyan, although it has been reported as common in the Fayoum. Status in Sinai uncertain.

Habitat: Generally found in sandy habitats with cover such as veg-etated dunes, tamarisk, palm, or acacia groves. Rarely recorded from hard ground. In the Fayoum, recorded from vegetated desert around springs.

Habits: Largely nocturnal though has been observed during the day. Colonial, living in complex burrow systems with escape tunnels and nest chambers, and with numerous openings generally dug into a vegetated sand mound or beneath dense vegetation. Usually burrows in sand or soft soil. When fleeing, runs for its burrow with its tail held erect, making the prominent black tip particularly conspicuous. Warning call a 'tick.' Diet mainly vegetarian including leaves and shoots. Has been recorded as storing food in the burrow. Gestation unknown. Breeds from April to May. Litter size 2–4.

Associated Species: The Silky Jird possibly around Sallum. In Saudi Arabia, recorded from same locations as Wagner's Gerbil and Baluchistan Gerbil.

Notes: In some of the literature, reference is made to *Meriones caudatus*. This species is not recognized, and all specimens relating to it were referred to the Libyan Jird *M. libycus* in Osborn and Helmy (1980) and followed here. See also notes on Shaw's Jird.

Similar species: Other jirds. In the hand, the Libyan Jird is the only Egyptian species with black claws and orange base to the tail. In the field, it is darker than the other species and the black tip to the tail is longer and more conspicuous than in other species except for the Negev Jird *Meriones sacramenti*, which has only been recorded in Egypt from northeastern Sinai.

SHAW'S JIRD *Meriones shawi* (Duvernon, 1842)
Pl. 26
Subspecies occurring in Egypt: *M. s. isis*.
Arabic: *Maryunaz shawi*
Identification: Length 250–315mm; Tail 122–155mm; Weight 70–120.2g. Large jird. Above yellow-brown, rather dark. Below, white with clear demarcation marked by a strip of yellow to orange-yellow along the flanks and onto the outer forelegs and hind legs. Underside with variable amounts of pale yellow. Feet white with narrow flank strip sometimes extending to them. Claws pale. Soles of hind feet partially haired. Head as upper parts, but

with gray spots above and below eye. Sparse hair on ears, with small, whitish patch at back. Tail long, paler than upper parts and not distinctly bicolored. Tuft black extending along terminal quarter of the tail. **Range and status:** North Africa from Morocco east to Egypt. In Egypt, only record-ed from the Mediterranean coastal desert from Sallum to

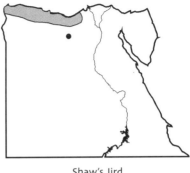

Shaw's Jird
(Meriones shawi)

Alexandria. Southernmost record from northeastern corner of Qattarra Depression.

Habitat: Variable within limited range having been recorded from sandy dune areas, coastal dunes, rocky areas, agricultural land, and vegetated clay areas.

Habits: Probably very similar to the Libyan jird, though burrows have been recorded more often in areas of hard, compacted soil. Diet probably similar but also including grain as it has been recorded from barley fields. Reproduction not known.

Notes: Harrison and Bates (1991) reject Shaw's Jird as a species and regard it as a possible subspecies of the Libyan Jird. This is sup-ported by the fact that the two have interbred in captivity and, though the Egyptian characteristics of the two jirds are distinct, elsewhere in the Middle East there is considerable confusion. In Egypt, the ranges barely overlap except at Sallum. The present description follows that of Osborn and Helmy (1980), though accepts Harrison and Bates' (1991) recommendation that the two species be reviewed.

Similar species: Other jirds. Distinguished from the Libyan Jird by the pale claws and less conspicuous tail tuft extending over only the terminal quarter. From Silky Jird, told by much smaller white patch behind the ear and some dark coloring on the ear.

TRISTRAM'S JIRD *Meriones tristrami* Thomas, 1892
Pl. 26

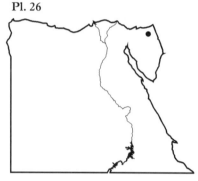

Tristram's Jird
(Meriones tristrami)

Subspecies occurring in Egypt: *M. t. tristrami*.

Identification: Length 246–274mm; Tail 125–140mm. A small, slender-limbed jird. Upper parts yellowish brown, pale to dark, with a narrow flank strip of yellow to orange extending onto outer fore-limbs and hind limbs. Fur short and dense. Underparts white. Feet white though flank strip may extend to them. Hind feet rather elongated. Soles of hind feet largely hairy. Claws pallid. Head as upper parts but paler particularly between eye and snout, and eye and ear. Distinct pale patch above each eye. Ears prominent and erect, almost naked and tip pigmented. White patch behind ear distinct. Tail yellowish brown above, orangish below at base and bicolored. Tail brush blackish but inconspicuous and only along terminal quarter of the tail.

Range and status: From Sinai through the Middle East north to Turkey and east to Iraq and Iran. In Egypt, only recorded from northeastern Sinai at al-Arish.

Habitat: Although only recorded from semi-arid coastal desert in Egypt, elsewhere has a wide habitat tolerance. Recorded from farm-land, meadows, steppe, sand dunes, and valley floors. It is not a jird of the desert proper and is not likely to be recorded far outside its very limited known range in Egypt.

Habits: Not known in Egypt. Elsewhere nocturnal and crepuscular. Burrows with several entrances in a variety of substrates but normally on slopes or earth mounds. Chambers with nest material found within burrow complex. Diet varied but mainly vegetable matter such as seeds, leaves, roots, grain, etc. Unlike other jirds, has not been record-ed as storing food in its burrow. Gestation 25–29 days. Litter size 1–7. Reported to be very prolific and may breed throughout the year.

Notes: Tristram's Jird is very variable in color and it has been suggested that the coloration of a given population is dependent on the ground color of the area it inhabits.

Similar species: Other jirds but distinguished from all by small size and inconspicuous tail tuft. Note that range in Egypt is very limited.

NEGEV JIRD (BUXTON'S JIRD) *Meriones sacramenti* Thomas, 1922

Pl. 26

Monotypic

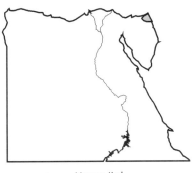

Negev Jird
(Meriones sacramenti)

Identification: Length 261–332mm; Tail 130–162mm. Large, dark jird with prominent ears and conspicuous tail tuft. Upper parts dark sandy-red, speckled black with a narrow cinnamon flank strip running along the side and extending onto the forelimbs and hind limbs and sometimes onto the foot. Underparts and inside of legs white. Feet white with soles of hind feet largely haired and claws whitish. Pale patch above eye indistinct. Ears prominent with the top third to tip pigmented. White hair on the inside, red-brown behind with small but clear white patch behind base of ear. Tail red-brown above with black hairs, slightly paler than upper parts, and paler below but not clearly bicolored. Tail brush conspicuous, black and extends along the terminal third of tail, often with some white hairs.

Range and status: Endemic to the Middle East. Recorded only from Israel, Palestine, and Sinai. In Egypt, known only from northeastern Sinai in the region of Rafah.

Habitat: Little known but thought to be restricted to sandy areas of coastal plains with stable or shifting dunes.

Habits: Nothing known in Egypt and though reportedly not uncommon within its very limited range in Israel, very little has been recorded about its behavior. Presumably nocturnal.

Associated Species: Recorded in the same habitat as the Greater Egyptian Gerbil and Anderson's Gerbil. Reported to compete with Tristram's Jird.

Similar species: Other jirds. Darker than the Libyan Jird with dark tip to the ear, darker, less orange flank stripe, buff (not orange) tail base and, in the hand, pale claws. Distinguished from the Silky Jird by dark tip to the ears and smaller white patch behind ears. The Silky Jird not yet recorded from Rafah. Both Tristram's and Shaw's Jird have inconspicuous tail brushes. Tristram's Jird is significantly smaller. Separated geographically from Shaw's Jird.

FAT-TAILED JIRD (FAT-TAILED GERBIL) *Pachyuromys duprasi* Lataste, 1880

Pl. 27

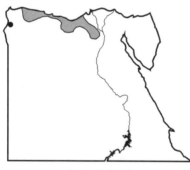

Fat-tailed Jird
(Pachyuromys duprasi)

Subspecies occurring in Egypt: *P. d. natronensis*.

Arabic: *Abu liyya*

Identification: Length 148–183mm; Tail 55–62mm; Weight 22–44.6g. Superficially similar to the jirds but with a distinctive, short, thick tail. Upper parts pale red-brown with black-tipped hairs running along the back. Fur long and fluffy. Flanks with clear, pale cinnamon strip, like colored strip in jirds, extending onto the hind limbs. Forelimbs, inside of hind limbs, and underparts white. No white rump patch. Ears fairly small, virtually naked and pigmented. White patch behind ear small. Pale spot above eye at best indistinct. Tail very distinctive. Shorter than head and body, club-shaped, and without terminal brush. Above pale red-brown, below white.

Range and status: North Africa north of the Sahara from Algeria to Egypt. In Egypt, found along the Mediterranean coastal desert, particularly the southern margin, from south of Sallum to the western margin of the Delta including Wadi Natrun and western margin of Nile Valley as far south as Abu Rawash.

Habitat: Sandy tracts with vegetation but also occasionally rocky areas. Seems to favor more sparsely vegetated areas than the relatively fertile Mediterranean coastal desert proper.

Habits: Nocturnal becoming active at dusk. In Egypt, burrow is simple, up to 1m deep, but elsewhere recorded as occupying more complex burrow systems, though these might not be dug by the Fat-tailed Jird. Diet in wild not fully known, though vegetable matter including seeds and plant stems probable, and possibly also insects. It has been suggested that it may also eat snails. Breeding habits unknown in the wild. Litters of 3–9 recorded in captivity between April and November.

Associated Species: Recorded from the same habitat as the Silky Jird, Lesser Egyptian Gerbil, Anderson's Gerbil, Pallid Gerbil, Greater Egyptian Gerbil, and Lesser Egyptian Jerboa.

Similar species: Although the Fat-tailed Jird is superficially similar to the true jirds, the tail shape is unique among Egyptian rodents and thus diagnostic.

FAT SAND RAT *Psammomys obesus* Cretzschmar, 1828
Pl. 27
Subspecies occurring in Egypt: *P. o. obesus*, *P. o. nicolli*, and *P. o. terrae-sanctae*.
Arabic: *Guradh*
Identification: Length 251–356mm; Tail 100–157mm; Weight 32–43g. Largest of the gerbil/jird group in Egypt. Variable but upper parts generally dark reddish brown speckled with black. Fur short and dense. Flanks are paler brown to yellowish extending onto the limbs. Underside and inner legs whitish, pale yellow to dark yellow. Chin and throat pale. Limbs powerfully built, digits with blackish claws. Feet buffish yellow above. Hind feet haired except for naked soles. Muzzle distinct. Ears small, rounded, and pigmented, with dense whitish to yellowish hair. White patch behind ear present but not conspicuous. No white patch above eye. Tail rather short and stout, well covered with short hairs and with a conspicuous black tip.
Range and status: North Africa from Morocco to Egypt, south to

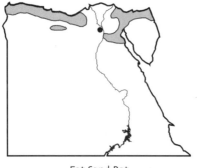

Fat Sand Rat
(Psammomys obesus)

Sudan and east to Sinai, Middle East, and parts of Arabia. In Egypt, restricted to the north of the country. *P. o. obesus* found along the Mediterranean coastal desert from Sallum to the west bank of the Rosetta branch of the Nile. Very few records from the Delta. *P. o. nicolli* recorded from Damietta east across the north coast and the shores of Lake Manzala to Port Said, and south to Ismailiya. *P. o. terraesanctae* recorded from the northern Eastern Desert and North Sinai at al-Arish to Rafah.

Habitat: Salt pans and salt marshes, sandy desert and wadi beds, and occasionally gravel pans. Found along roadsides and railway embankments. Distribution dependent on the availability of succulent plants on which it feeds and depends for moisture.

Habits: Unlike most desert rodents, the Fat Sand Rat is largely diurnal, though also comes out at night. It is by far the most likely of the jird/gerbil species to be seen by day foraging for food and even climbing small bushes. Will go out foraging, find a food or nest item, and dash back to the burrow with it to emerge soon after. Colonial living in an extensive but shallow burrow system dug beneath a shrub or pile of stones, etc. Studies in Algeria indicate that male territories can overlap and that territory size increases as food availability decreases. Active burrows often show accumulations of waste, networks of tracks, and droppings (Fat Sand Rats cover their scat like the jirds). Burrow systems have several openings. Nest chamber lined with dried vegetation and often human-made waste. Not thought to store food. Diet largely succulent desert plants. Food is eaten with one hand and wiped before consumption. When alerted to danger, will sit up in a manner similar to Prairie Dogs *Cynomys* spp. Predators include diurnal birds of prey and probably snakes. Gestation 23–25 days. Breeding in Egypt from September to May. Litter size 1–8.

Associated Species: The Fat Sand Rat shares the same habitat as many other jirds/gerbils. The Greater Egyptian Jerboa, Giza Gerbil, Pygmy Gerbil, House Mouse, and Long-eared Hedgehog *Hemiechinus auritus* have been recorded from Fat Sand Rat burrows as have cobras and scorpions.

Notes: The Fat Sand Rat is extremely variable in size and color, the latter often depending on habitat.

Similar species: Within its range, no other species of jird/gerbil is likely to be active by day. Can be told from the jirds by large size and heavy build, much smaller ears, yellowish belly, and uniform tail (not bicolored).

BUSHY-TAILED JIRD (BUSHY-TAILED DIPODIL) *Sekeetamys calurus* (Thomas, 1892)
Pl. 22

Subspecies occurring in Egypt: *S. c. calurus* and *S. c. makrami*.

Arabic: *Rishi al-dhayl*

Identification: Length 229–292mm; Tail 131–164mm; Weight 26.6–49.8g. A large, very distinctive jird with a long, bushy tail. Above, yellow-brown, speckled black as hair tips are black. Flanks paler and more clearly yellow to orange showing flank stripe as in the jirds) extending to

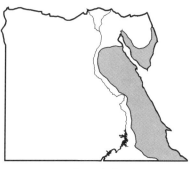

Bushy-tailed Jird
(Sekeetamys calurus)

the limbs. Inner limbs white. Below white, clearly demarcated. Pale round eyes though no distinct patch. Ears large and slightly elongated. Pigmented. Limbs slender with hind feet elongated. Soles of feet naked and pigmented, claws cream colored. No white rump patch. Tail long, long-haired, and bushy throughout its length. Largely grayish to black except at base where yellow-brown. Tip normally white. The tail is normally held erect and curving, never dragging the ground.

Range and status: Egypt and Sinai to Israel, Palestine, and Jordan.

Bushy-tailed Jird
(*Sekeetamys calurus*)

Also central Arabia. In Egypt, *S.c. makrami* is found throughout the Eastern Desert from Wadi Hof and the cliffs at Ain Sukhna south to Gebel Elba. *S.c. calurus* is found along the western ranges of Sinai and the mountains of South Sinai including around St. Katherine's Monastery. Reportedly common in South Sinai.

Habitat: Restricted to rocky desert mountain ranges, cliff slopes, screes and mountain tops. Also reported from buildings in these regions.

Habits: Nocturnal. Burrow or a den amongst boulders and beneath rocks. Emerges at dusk to feed amongst the rocks, being extremely agile. Feeds on vegetable matter including green vegetation and dry seeds. May also eat insects if available. Breeding season unknown in wild; in captivity, it breeds throughout year.

Associated Species: The Middle Eastern Dormouse, Cairo Spiny Mouse, Golden Spiny Mouse, and Wagner's Gerbil are found in similar habitat to the Bushy-tailed Jird.

Similar species: The bushy, dark tail of the Bushy-tailed Jird is unique amongst Egyptian rodents, except for the Middle Eastern Dormouse. This species, however, has a clear and striking facial mask and does not carry its tail aloft like the Bushy-tailed Jird often does.

The Jerboas—Family Dipodidae
30 species worldwide with 3 in Egypt.

The jerboas bear a superficial resemblance to the gerbils and are equally well-adapted to life in the desert, sharing the latter's nocturnal lifestyle, spending the day in a burrow, and obtaining moisture from their food, mainly seeds, but also other vegetation and insects. They are, however, quite distinct and notable, mainly, for the enormous length of the hind limbs that may be four times longer than the forelimbs. It is not surprising that the Jerboas are jumpers, bounding along on their hind legs using their long and prominently tufted tail for balance. At rest, too, they are bipedal, the tail here being used as a prop like a kangaroo. So that the comparatively huge feet can handle the stresses of this bounding locomotion, the three main bones of the hind feet are fused to form a single bone known as the 'cannon' bone. The forelimbs are used primarily for feeding and grooming.

As in the gerbils, the eyes and ears are both prominent, an adaptation to nocturnal life, but this is especially marked in the rare Four-toed Jerboa where the ears are almost rabbit-like in shape and proportion. In the two *Jaculus* species, there is a fold of skin that can be extended over the nostrils to keep sand out during burrowing.

As a family, the Jerboas are relatively easy to identify in the field due to their bipedal stance and distinctive bounding gait. They are most often seen at night in the car headlights bounding across the desert or across desert roads. They may freeze, rabbit-fashion, in headlights. Within the family, species identification is more difficult though the only widespread Jerboa species in Egypt is the Lesser Egyptian Jerboa. Most care with identification should be taken in the Mediterranean coastal desert, where all three species occur, and in Sinai, where the two *Jaculus* species overlap.

FOUR-TOED JERBOA *Allactaga tetradactyla* (Lichtenstein, 1823)
Pl. 22
Monotypic
Arabic: *Ghufl*
Identification: Length 256–299mm; Tail 154–180mm; Weight 48–56g. Male slightly larger than female. A small but distinctive long-eared, long-tailed jerboa. Upper parts speckled black-and-orange due to black and orange banding of the hairs. Pales to gray along the sides with orange rump. Underparts and forelimbs white. Hip bands white, not meeting at tail base. Hind limbs powerful and greatly elongated, white with back of lower leg black. Hind feet greatly elongated for hopping gait. Three toes plus a vestigial fourth, unique amongst Egyptian jerboas. Snout buffish, white patch below eye. Ears long, slender, and erect like a hare's, but with

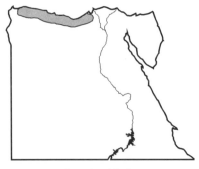

Four-toed Jerboa
(Allactaga tetradactyla)

rounded tips. Ears pigmented and with white hairs. Tail long, paler than upper parts above, whitish below. Feathered tip black with white at the very tip.

Range and status: Restricted distribution in eastern Libya to Egypt. In Egypt, reported from the Mediterranean coastal desert extending almost as far east as the western margin of the Delta at Burg al-Arab. As early as the 1960s, this species was reported to be locally extinct and endangered by habitat destruction through land reclamation. Tourist development along the coast is also probably contributing to its decline.

Habitat: In and around salt marshes extending inland to cultivated areas such as barley fields.

Habits: A little known species. Nocturnal, spending the day in a simple but deep burrow. May use the burrow of the Greater Egyptian Jerboa. Diet and breeding not known.

Associated Species: Occurs in the same salt marsh habitat as the Fat Sand Rat and the Greater Egyptian Jerboa. Inland also found with Shaw's Jird.

Similar species: Other Egyptian jerboas from which it may be distinguished by the much longer, more slender ears, and darker color. In the hand, the vestigial fourth toe is diagnostic. From the Greater Egyptian Jerboa further told by much smaller size.

Four-toed Jerboa
(Allactaga tetradactyla)

LESSER EGYPTIAN JERBOA (LESSER JERBOA) *Jaculus jaculus* (Linnaeus, 1758)

Pl. 22

Subspecies occurring in Egypt: *J. j. jaculus*, *J. j. flavillus*, *J. j. schlueteri*, and *J. j. butleri*.

Arabic: *Yarbu' hurr, Abu nawara*

Identification: Length 245–325mm; Tail 150–205mm; Weight 43–66.4g. Small, rather variable jerboa. Upper parts varying in color from orangish to reddish brown. Fur long and soft. Flanks orange to gray. Thighs orange

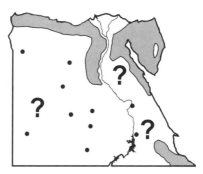

Lesser Egyptian Jerboa
(Jaculus jaculus)

to reddish brown with pale, gray to white thigh bands meeting at base of tail. There is some evidence that desert animals are brighter colored than specimens from non-arid areas. Underparts and forelimbs white. Hind limbs elongated, white, and may or may not be black at the back of the lower leg. Hind feet greatly elongated with three toes, the central one being longest. Sole of hind foot naked with toe pads obscured by long, silky white fur that probably gives the jerboa purchase on soft sand. Head proportionately large with prominent eyes, and large, rounded ears, not highly elongated and only pigmented at tip. Face largely pale. White patch above eye prominent. White patch behind ear indistinct. Tail long, colored same as the upper parts

above, paler below. Coloration of feathered tip complex and tricolored—white, black, white. Terminal tip white preceded by sub terminal black band that may or may not be white below. This is, in turn, preceded by a whitish band.

Range and status: North Africa from Morocco south to Mauritania and east through Chad and Niger to Somalia, north to Egypt. Also Sinai east to Middle East, Arabia, Iraq, and Iran. In Egypt, widely distributed but never common. *J. j. jaculus* found throughout the Western Desert to the western margin of the Nile Valley and north to the margins of the Mediterranean coastal desert. Recorded from environs of Cairo including along the Cairo–Alexandria Road and near the pyramids. Also noted from all the major Western Desert oases including Siwa, Wadi Natrun and the Fayoum, and Wadi al-Rayyan, where reportedly rare. *J. j. flavillus* recorded from the Mediterranean coastal desert from Sallum to the western margin of the Delta and south to Wadi Natrun. *J. j. schlueteri* recorded from the Eastern Desert from the eastern margin of the Delta south as far as Luxor and Qena. There seems to be a particularly large number of records from around the Cairo–Suez Road. Also recorded from Wadi Digla. *J. j. butleri* recorded from the southern Eastern Desert from Aswan south, including the Gebel Elba region.

Habitat: Has a high habitat tolerance and regarded by Harrison and Bates (1991) as "one of the most successful mammalian colonists of the desert peninsula of Arabia." In Egypt, recorded from sand dune habitat, plains, wadi floors, etc. In Sinai, recorded as high as 1,500m in St. Katherine. In Gebel Elba, found in open acacia woodland and in the Mediterranean coastal desert, recorded from habitat in and around cultivated areas to barren desert further inland. In the Fayoum, recorded from vegetated areas at desert margins around springs.

Habits: Nocturnal and generally active throughout the night. Spends the day in a simple burrow excavated in hard substrate a meter or more deep. The burrow is plugged in summer, unplugged in winter. There may be one or more escape tunnels but these are plugged. Nest of dry plant material generally in the lowest part of the burrow. Less sociable than Greater Egyptian Jerboa and more likely to be found alone. May wander extensively. Vegetarian diet, feeding on plant stems, seeds, grain, leaves, etc., depending on habitat. Apparently does need to drink but can get enough moisture from dew without requiring standing water. Predators include most desert carnivores, birds of prey, large lizards, and humans. A disturbed Lesser Egyptian Jerboa will head for its burrow. Escape flight erratic and may also try to elude predators by remaining motionless. Gestation about 25 days. In Egypt, breeding probably from February to September. Litter size 4–10.

Associated Species: Because of its wide range and high habitat tolerance, the Lesser Egyptian Jerboa has been found with a wide range of other species including the Four-toed Jerboa, Greater Egyptian Jerboa, Libyan Jird, Shaw's Jird, Silky Jird, Lesser Egyptian Gerbil, Anderson's Gerbil, Greater Egyptian Gerbil, Pallid Gerbil, and Fat-tailed Gerbil.

Notes: The most widespread and common jerboa in Egypt.

Similar species: Other jerboas. See previous species to distinguish from the Four-toed Jerboa. From Greater Egyptian Jerboa, told by much smaller size, paler color, shorter ear and hind foot, and ears pigmented only at tip. Much more widespread.

GREATER EGYPTIAN JERBOA *Jaculus orientalis* Erxleben, 1777
Pl. 22

Subspecies occurring in Egypt: *J. o. orientalis.*

Arabic: *Qirifti*

Identification: Length 332–403mm; Tail 195–243mm; Weight 108.6–147g. A large jerboa, more than twice the weight of the Lesser Egyptian Jerboa. Upper parts orangish brown, fading to gray on the sides, and orange to buff along the flanks. Coat long, soft, and dense. Hind limb

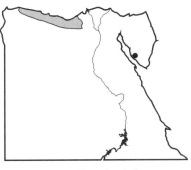

Greater Egyptian Jerboa
(Jaculus orientalis)

brownish orange on the outside with hip band grayish meeting at base of tail. Underparts and forelimbs white. Inner hind limbs white.

Greater
Egyptian Jerboa
(Jaculus orientalis)

Hind feet enormous, white above, with three toes, the central being the longest. Claws brownish. Forefeet have four toes, the first of which is poorly developed. Head proportionately large, mostly buffish with poorly defined pale spots above the eye and behind the ears. Ears large, longer than in Lesser Egyptian Jerboa and broader than in the Four-toed Gerbil. Fully pigmented. Whiskers strikingly long, reaching over 8cm. Tail long, colored as upper parts above, whitish below, with bushy terminal tuft that is tricolored—cream, black, white.

Range and status: North Africa from Morocco east to Egypt, the Sinai, and the Negev Desert, Israel. In Egypt, recorded from the Mediterranean coastal desert from Mersa Matruh to Alexandria, and from North Sinai at Nakhl and Gebel Maghara, with a single record from southwestern Sinai. Much more local than the Lesser Egyptian Jerboa.

Habitat: In Sinai, from seashore along the Gulf of Suez coast and from sandy desert in the north. In the Mediterranean coastal desert, in and around salt marshes, sandy and rocky slopes, and cultivated areas, including barley fields and olive groves.

Habits: Nocturnal but may also be active at dusk. Sociable, rarely encountered alone. Burrows generally in hard substrate up to about 2m long and plugged during the summer, left open during the winter. There may be escape tunnels. Evidence suggests that the Greater Egyptian Jerboa prefers hillside locations in winter and being near arable areas during the summer. Nest chamber may contain animal fur as bedding. Diet includes roots, shoots, and seeds, including grain, and probably succulents. May store food in small chambers off the main burrow. Probably obtains enough moisture from food (especially fresh shoots and succulents) and does not need to drink. Probably several pregnancies during the year. Gestation unknown. Breeding recorded from November to July. Litter size 2–5.

Associated Species: The Greater Egyptian Jerboa has been recorded in the same habitat as the Four-toed Jerboa, the Fat Sand Rat, Shaw's Jird, Anderson's Gerbil, and the Lesser Egyptian Jerboa. It has been recorded as sharing burrows with the Fat Sand Rat and the Four-toed Jerboa has been found in Greater Egyptian Jerboa burrows. On field margins, it has been found in the burrows of Shaw's Jird.

Notes: Significantly larger than the Lesser Egyptian Jerboa.

Similar species: Other jerboas, see previous two species.

The Mouse-like Rodents—Family Muridae (in part)
The Old World Rats and Mice—Subfamily Murinae
408 species worldwide with 7 species in Egypt.

The relative paucity of murine species in Egypt is emphasized further when one considers the fact that three of these species, the House Mouse, and the Brown and House Rats, are introduced and a further, the Bandicoot Rat *Nesokia indica*, is considered relict. This paucity is probably because the murines are generalists and, large though the subfamily is, very few have become adapted to life in the desert. The seven species found in Egypt are largely confined to the Nile Valley, Delta, and areas such as oases, where water and human settlements exist. The two exceptions are the Cairo and Golden Spiny Mice but even these are absent from truly arid desert.

What they lack in representation, though, they make up for in success, largely because of their ability to live with—even rely on—humans and their activities. The House Mouse, Brown Rat, House Rat, and Cairo Spiny Mouse have all adapted to living in the midst of human habitation, even in the center of Cairo. The Nile Kusu *Arvicanthis niloticus* is a common inhabitant and serious agricultural pest in the Delta and Nile Valley. Only the Golden Spiny Mouse and the Bandicoot Rat show more restricted distributions. The Golden Spiny Mouse can be found in genuinely wild desert wadis but may also be seen amongst the garbage and tea shops at the summit of Mount Sinai. The Bandicoot Rat is localized in the western Delta and adjoining oases, restricted to cultivated areas.

The four larger murines can readily be told from other mouse-like rodents by their large size, but also by their coarse coats, thick tails, and 'normally' proportioned hind limbs. The two spiny mice can be distinguished from the superficially similar gerbils by the spiny coat, much larger ears, and, in the case of the Golden Spiny Mouse, the color. Color and form also help distinguish the House Mouse from the gerbils, along with the absence of a terminal tuft to the tail.

CAIRO SPINY MOUSE (EGYPTIAN SPINY MOUSE) *Acomys cahirinus* (Desmarest, 1819)

Pl. 28

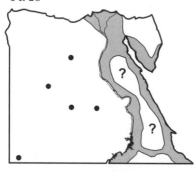

Cairo Spiny Mouse
(*Acomys cahirinus*)

Subspecies occurring in Egypt: *A. c. cahirinus*, *A. c. dimidiatus*, *A. c. megalodus*, *A. c. hunteri*, *A. c. helmyi*, and *A. c. viator*.

Arabic: *Fa'r abu shawk qahiri*

Identification: Length 160–276mm; Tail 85–138mm; Weight 20.9–64g. Small- to medium-sized mouse. Color very variable ranging from uniform dark gray with white feet in Delta and Nile Valley to pale brown above, tinged orange along flanks, white below with pale limbs and white feet in desert populations, with many variations in between. Palest in Sinai form *A. c. dimidiatus*. Fur from behind the shoulder along back to tail base spiny. In most forms, distinct white patch below eye, clearly visible in field, and behind ears. Rather pointed snout with clear delineation of white underside. Ears large, whitish. Eyes prominent. Tail about length of body, brownish above, pale below, sparsely haired with scales distinct.

Range and status: North Africa from Mauritania to Egypt, south to Nigeria, east to Tanzania. Sinai, Middle East, and north to Turkey, with populations on Cyprus and Crete. Also Arabia and east to Iran and Pakistan. In Egypt, the common mouse over much of the country. *A. c. cahirinus* in the Delta, including Alexandria, and Nile Valley from Cairo south to Aswan. In Western Desert, in Bahariya. *A. c. dimidiatus* common in South and western Sinai, including St. Katherine and Wadi Feiran. Also North Sinai to south and southeast of al-Arish. *A. c. megalodus* in Eastern Desert north of Wadi Araba, including Wadi Rish Rash. *A. c. hunteri* in Eastern Desert south of Wadi Araba to shores of Lake Nasser and Gebel Elba region. *A. c. helmyi* in Western Desert oases of Farafra, Dakhla, and Kharga. *A. c. viator* confirmed only from Gebel Uweinat.

Habitat: Found in a wide range of habitats. In Delta and Nile Valley often in human settlements, including towns and cities. According to Osborn and Helmy (1980), it is dominant over the House Mouse in these areas where it is virtually commensal. In villages, tombs, grain stores, etc. In desert, found in rocky wadis, cliff sides, and palm groves. In Wadi Rishrash, the Cairo Spiny Mouse was found on the wadi floor while the Golden Spiny Mouse was found up boulder-strewn canyons cut through wadi cliffs. In South Sinai, in rocky habitat to 1,800m, above which replaced by the Golden Spiny Mouse.

Habits: Active by day and night, though probably predominantly nocturnal. Burrow in sand or gravel, or in crevices amongst boulders, and in houses. Wide variety of food recorded from plant stuff, seeds, dates, and grain, to snails, insects, including moths and grasshoppers, spiders, and scorpions. Will also eat dung. Though the Cairo Spiny Mouse can survive without water, it can only do so if moist food, e.g. snails, is available. Predators include most carnivores as well as birds of prey, and owls, and probably snakes. Defense strategies include biting and erecting spines, making the animal appear bigger than it actually is. Can also shed its tail and many Cairo Spiny Mice may have shorter tails than measurements indicate or may be completely tailless. Gestation about 42 days. Breeding

recorded throughout year but, as it is sensitive to cold, breeding probably less likely in winter. Litter size 1–6 with average of 3.

Associated Species: Found with a wide range of other species including the House Mouse and Golden Spiny Mouse discussed above. Also with Middle Eastern Dormouse (in Sinai only), Bushy-tailed Jird, Wagner's Gerbil, Lesser Egyptian Gerbil, and Large North African Gerbil. In urban areas may share habitat with both House and Brown Rats.

Notes: Care should be taken in handling the Cairo Spiny Mouse. Not only does it bite, but certain populations have been found to carry organisms responsible for typhus, tick typhus, and Q-fever.

Similar species: Other mice. The House Mouse is smaller and paler than the commensal Cairo Spiny Mouse with no spines on its back. The Golden Spiny Mouse has very different coloring and proportionately shorter tail, more extensive spines, and smaller ears. Also, the Golden Spiny Mouse is more strictly diurnal and has a much more restricted range. Rat species are all much larger and heavier.

GOLDEN SPINY MOUSE *Acomys russatus* (Wagner, 1840)
Pl. 28

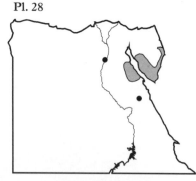

Golden Spiny Mouse
(*Acomys russatus*)

Subspecies occurring in Egypt: *A. r. russatus* and *A. r. aegyptiacus*.

Arabic: *Fa'r abu shawk dhahabi*
Identification: Length 146–203mm; Tail 56–81mm; Weight 24–53.2g. Small but stocky mouse. Upper parts golden-orange to reddish, flanks yellow, and underside pale. Fur spiny from head to base of tail, i.e., spines much more extensive than in the Cairo Spiny Mouse. Legs gray. Feet pale with black soles. Head as upper parts. White spot below eye small but distinct. Ears proportionately smaller than previous species, black but with pale fur and

white patch behind each ear. Tail noticeably shorter than body (but beware of Cairo Spiny Mice with shed tails) and blackish above and below.

Range and status: Egypt's Eastern Desert, South Sinai, Israel and Palestine, and Arabia. In Egypt, *A. r. russatus* found in South Sinai, including St. Katherine, where it has even been recorded on the summit of Gebel Musa. Less common than the Cairo Spiny Mouse where ranges overlap. *A. r. aegyptiacus* in northern Eastern Desert including Wadi Rishrash. Isolated record from as far south as Wadi Atalla. Probably not uncommon within limited range and habitat.

Habitat: More strictly a mouse of arid, rocky areas than the Cairo Spiny Mouse. Recorded from 2,286m on Gebel Musa whereas previous species, with less tolerance of cold, only found to 1,800m.

Habits: Unlike the Cairo Spiny Mouse, largely diurnal and one of the few desert rodents that can be seen by day. Does not burrow but lives in rock crevices, amongst boulders, etc. May enter buildings in St. Katherine. Can be seen on the hike up to Gebel Musa and on the summit can be found foraging amongst garbage. Diet, predators, and escape strategies probably similar to the Cairo Spiny Mouse, though reportedly less aggressive. Gestation 44 days. Female builds nest of dry vegetation for young. In Egypt, breeding probably spring to autumn. Litter size 1–4.

Associated Species: Found with Cairo Spiny Mouse, Bushy-tailed Jird, Wagner's Gerbil, and Middle Eastern Dormouse (not in Eastern Desert).

Notes: Beware: mice with shed tails can look very much like hamsters!

Similar species: Same as the Cairo Spiny Mouse.

HOUSE MOUSE *Mus musculus* Linnaeus, 1766
Pl. 28

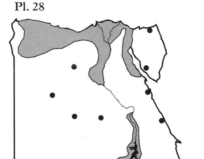

House Mouse
(Mus musculus)

Subspecies occurring in Egypt: uncertain.
Arabic: *Fa'r al-manzil*
Identification: Length 108–200mm; Tail 53–97mm; Weight 9.4–20.9g. Small mouse and the only Egyptian mouse with no spines in the fur. Highly variable in color from grayish through beige to light or dark brown. Flanks paler. Underside whitish, often tinged buff though may be dark. Shade seems to be related to habitat. Head with pointed muzzle, indistinctly marked (unlike spiny mice) with indistinct paler spot below eye and behind ears. Ears proportionately smaller than in Cairo Spiny Mouse. Tail as upper parts above, paler below, equal to or slightly longer than body. 150–200 tail rings can be seen beneath fur. No terminal tuft.

Range and status: Worldwide distribution, thanks to humans, but originally probably from Turkmenistan region. In Egypt, widely distributed along the north coast including North Sinai. In the Western Desert, found in all the main oases including Siwa. The Delta and Wadi Natrun south to Cairo, the Nile Valley, including the Fayoum, where reportedly very common, south to Aswan and Abu Simbel. Absent from Eastern Desert but found in settlements along Red Sea coast south to Marsa Alam. Port Said and Suez with isolated records in Sinai from al-Tor and al-Arish. Expansion of range likely as human settlement spreads.

Habitat: Inhabits a wide range of habitat from agricultural areas, villages, towns, cities to coastal desert. Also beaches, seashore, salt marshes, meadows, canal banks, and palm and olive groves. Absent only from true desert but often present at oases where probably transported by humans. Where it occurs with the Cairo Spiny Mouse, it is the latter that seems dominant, becoming the true 'house' mouse—living inside human settlements.

Habits: Generally nocturnal but may be active during the day. Usually lives in a sheltered, shallow burrow, self excavated, with nest lining. Communal. Diet is vegetarian and can be a pest in grain stores. Predators as for the previous species. Wary and secretive. Voice the oft-imitated squeak. Gestation about 22 days. Probably breeds throughout the year. Individual litter size up to 10 or more, but communal litters may consist of many more young.

Associated Species: Due to its very wide distribution, the House Mouse has been shown to associate with most other Egyptian rodents except, perhaps, true desert species.

Similar species: Spiny mice have distinctly spiny fur and are larger. The Golden Spiny Mouse is differently colored. The Cairo Spiny Mouse is very variable but usually has distinct facial markings and larger ears. Gerbils often have a terminal tuft to tail and more highly developed hind limbs.

NILE KUSU (AFRICAN GRASS RAT, GRASS RAT, FIELD RAT, KUSU RAT)
Arvicanthis niloticus (Desmarest, 1822)
Pl. 29

Subspecies occurring in Egypt:
A. n. niloticus.
Arabic: *Fa'r al-ghayt*
Identification: Length 284–375mm; Tail 125–173mm; Weight 102–201.2g. Large, big-headed, rather slim rat. Upper parts rather grizzled-black and yellow. Flanks paler. Coat coarse. Dark stripe down center of back may be distinct running from top of the head to the base of

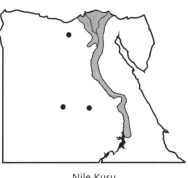

Nile Kusu
(Arvicanthis niloticus)

the tail. Underside whitish to grayish. Hairs on rump longer and more or less tinged yellow to orange. Feet buffish orange to blackish above. Head large, rather pointed. Whiskers sparse and short. Ears small and rounded, tinged orange. The tail is about 80% of body length, slender, blackish above, pale below. Fur dense, concealing tail rings. No terminal tuft.

Range and status: Northeastern Africa from Egypt and Sudan south to Kenya and Tanzania west to Uganda, Nigeria, and Senegal. Also southwestern Arabia. In Egypt, largely confined to the Delta and Nile Valley as far south as Aswan. Also Kharga and Dakhla, Wadi Natrun and the Fayoum, where reportedly common. Generally common in agricultural areas and is an important pest species.

Habitat: Agricultural areas and margins, canal and railway embankments with good cover, olive groves, and gardens. Recorded around, but apparently never in, human settlements. Presence in oases probably due to humans. Species does not seem to be able to compete with the Brown Rat in settled areas. By 1910, the Brown Rat, reportedly, had totally ousted the Nile Kusu from the Giza Zoo in a period of only 8 years. Not a desert animal.

Habits: Active by day and night. Digs a long, sinuous, rather shallow burrow, complex with many entrances. Shorter burrows used for temporary shelter. Openings not plugged. Generally solitary or in pairs, though larger groups may congregate in burrows. Diet mainly grain, but also takes fruit, bark, etc. Predators include the Egyptian Mongoose *Herpestes ichneumon*, Swamp Cat *Felis chaus*, probably the Red Fox *Vulpes vulpes*, birds of prey, and snakes. Decimation of these predators thought to be responsible for its current large numbers. Gestation period about 20 days. Main breeding period thought to be June to November but may breed at other times. Litter size 5–6, possibly up to 10, and there may be up to 4 litters per year.

Associated Species: Some overlap occurs with the Cairo Spiny Mouse, House Mouse, and the House and Brown Rats.

Similar species: Other rats. Easily distinguished from both the House and Brown Rat by distinctive color and well-furred tail. Much shorter tail and smaller ears than the Black Rat and much smaller than the Brown Rat. Distinguished from the Bandicoot Rat by well-furred tail, coarser coat, and orange patch at base of tail. The Bandicoot Rat rarely seen out of its burrow. A short-tailed rat seen in the open is far more likely to be the Nile Kusu.

HOUSE RAT (SHIP RAT, BLACK RAT, ROOF RAT) *Rattus rattus*
(Linnaeus, 1758)
Pl. 29

Formerly divided into as
many as 4 different sub-
species in the region but these
were based largely on color;
however, and as two or more
phases may occur together
(and there are many interme-
diates), these subspecies are
now generally not recognized.

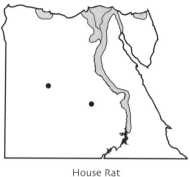

Arabic: *Guradh manzil,
Guradh aswad*

House Rat
(Rattus rattus)

Identification: Length 344–
452mm; Tail 188–244mm;
Weight 87–174g. Large, long-tailed, big-eared rat. Despite name,
coat variable and rarely black. Upper parts vary from almost black
through dark brown to pale brown, flanks paler and underparts from
grayish to white. Hair coarse, especially along back. Feet gray, buffish
to whitish. Head rather pointed with prominent eyes. Ears large,
rounded, and almost naked. Tail significantly longer than head and
body, and slender. Virtually naked, with sparse stiff hairs and many
(generally over 200) rather indistinct rings. No terminal tuft. Tail
used for balance when climbing and held off ground when running.
Voice includes various squeaks and screams.

Range and status: Worldwide distribution thanks to humans probably originating from India. Some populations, e.g., in the U.K., are
declining due to competition with Brown Rat and may even be locally endangered. In Egypt, found throughout the Delta and Nile Valley
south to Abu Simbel, and recorded in towns elsewhere, including
Mersa Matruh, al-Arish, Fayed, Suez, and the oases of Wadi Natrun,
Fayoum, Kharga, and Farafra. Common in Cairo and reportedly
very common in the Fayoum. It is uncertain whether the House Rat
is declining in Egypt due to competition with the Brown Rat.

Habitat: Commensal occurring in cities, towns, and villages. Also
agricultural areas, canal and railway embankments, and gardens.

Habits: Active by day but probably more so by night. Digs a shallow burrow, often with many openings that are unplugged, or uses burrow of other species or natural crevices. Will also build nest of grass and twigs in a palm tree. Often found in upper stories of buildings or granaries. Diet is vegetarian including grain, fruit, and seeds. Agile and climbs well, including trees where it will climb to feed on seeds, e.g., Kurrajong *Brachychiton populneus* seeds. Less fond of water than Brown Rat but can and does swim. Elsewhere in region, fish and water snails included in diet. Generally wary but can sometimes be seen in open in broad daylight. Gestation 21 days. Probably breeds all year round producing up to 5 litters a year of 1–8 young.

Associated Species: Humans. As a commensal, the House Rat is frequently found in the same habitat as the Brown Rat, House Mouse, Cairo Spiny Mouse, and, to a lesser extent, the Nile Kusu.

Similar species: See Nile Kusu. Distinguished from the Brown Rat by smaller size, more slender build, larger ears, and longer tail (which is longer than body). From the Bandicoot Rat, told by much longer tail, more pointed snout, and very different behavior. Both the House and Brown Rats can look disconcertingly like weasels at night when they scuttle across the street. However, look for the much longer tail even in the Brown Rat, and more robust body.

BROWN RAT (NORWAY RAT, COMMON RAT, SEWER RAT) *Rattus norvegicus* (Berkenhout, 1769)
Pl. 29
Probably monotypic.
Arabic: *Guradh nurwigi, Guradh bunni*
Identification: Length 341–488mm; Tail 145–234mm; Weight 208.3–360g. Large, thickset rat. Dark brown above with flanks tinged grayish and underparts grayish white. Overall, rather dull and uniform. Fur coarse. Feet whitish, sparsely haired. Head with less prominent eyes than previous species and blunter snout. Ears smaller. Tail is generally shorter than head and body, much thicker than that of the House Rat, and virtually naked, with generally under 200 rings. No terminal tuft. Tail held clear of the ground when running. Voice various squeaks, grunts, and screams.

Range and status: World-wide distribution thanks to humans, probably originating around the Caspian Sea region. In Egypt, restricted to the Delta and Nile Valley as far south as Aswan. Also Ismailiya and Suez, with one isolated record from the Gebel Elba area. Less common and widespread in Egypt than the House Rat. Common in the Fayoum.

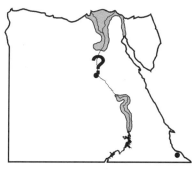

Brown Rat
(Rattus norvegicus)

Habitat: As for the House Rat, but seems more inclined to wetter habitats, including the seashore.

Habits: Active by day but far more often at dusk and at night. Burrow is shallow, complex with several openings that are unplugged. Burrow situated below buildings in canal banks, etc. Social. Diet very varied including grain, fruit and vegetables, refuse, eggs, meat, and may kill small birds and mammals. Even recorded having killed House Rat. Less agile than the House Rat and much less likely to climb trees. Swims well and readily takes to water. Gestation 21 days. Elsewhere in region recorded as having up to 5 litters a year, of 2–12 young.

Associated Species: Humans. See also House Rat.

Similar species: See Nile Kusu and House Rat. Much larger and has a longer tail than the Bandicoot Rat, which is very rarely seen in the open and has a much more restricted range. Also no white patch on throat. Like the House Rat, may be confused with a weasel at night, but more robust body and longer tail.

BANDICOOT RAT (PEST RAT, SHORT-TAILED BANDICOOT RAT) *Nesokia indica* (Gray and Hardwicke, 1830)

Pl. 29

Subspecies occurring in Egypt: *N. i. suilla.*

Arabic: *Abu 'afan*

Identification: Length 275–331mm; Tail 110–134mm; Weight 205.5–280g. Large, thickset, rather short-tailed rat. Above brown,

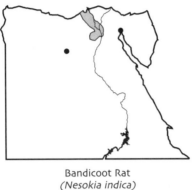

Bandicoot Rat
(Nesokia indica)

buffish along flanks and underside. Coat short, dense and soft unlike *Rattus* rats and Nile Kusu. Inside of limbs and feet white. Claws long. Head proportionately large and heavy with white patch on throat, though this may be small or sometimes absent. Snout short. Ears proportionately large and rather long, almost naked. Pale patch at back of base of ear may be present. Tail thick and shorter than head and body. Dark and sparsely haired. No terminal tuft.

Range and status: Egypt east to Israel, Syria, southern Turkey and Asia Minor, Iran, Afghanistan, and India. Egypt's population is regarded as relict, isolated from the Asiatic population. Largely confined to the western part of the Delta south to Cairo, Wadi Natrun, Fayoum (where common), and Bahariya. There is a small population in Suez of uncertain origin. Absent from Sinai, but found in Israel close to the Israeli–Egyptian border.

Habitat: Always found in damp habitats such as well-irrigated farm land, canal banks, irrigation ditches, lake shores, etc. In the Fayoum also recorded around settlements.

Habits: Probably largely nocturnal but has been seen by day, possibly due to burrow flooding. Basically very rarely seen outside its burrow. Burrows are extensive and complex, up to a reported 9m in length, but not deep. Burrow system includes a nest chamber and numerous side burrows and entrances, some of which are plugged. Heaps of soil may be present outside unplugged entrances. Diet is vegetarian, feeding largely on roots, grain, fruit, including water melon, and vegetables. Predators include snakes, owls, and the Swamp Cat. Reportedly defends itself vigorously by biting. Elsewhere only one Bandicoot Rat has been found per burrow. Gestation period 26–28 days. Probably breeds throughout year with litter size of 1–8.

Associated Species: The Greater Egyptian Gerbil has been found in the burrow of the Bandicoot Rat. Habitat may overlap with other rats.

Notes: It is interesting that, despite living an almost totally underground existence (as far as is known), the Bandicoot Rat shows very little adaptation to this, except for its long claws. Compare with the Lesser Molerat.

Similar species: See other rats. This species is very rarely seen above ground and its burrowings are the best sign of its presence in a suitably damp habitat.

The Dormice—Family Gliridae

10 species worldwide with 1 species in Egypt.

The dormice form a small, but distinctive, family of rodents. In form and habits, they are somewhat intermediate between the true mice and the squirrels: they are mouse-like in form but are stockier, have a bushy tail, and are clearly adapted for climbing. Such adaptations include naked soled feet, short, curved claws (particularly on the hind feet), and a long tail that probably helps with balance. The single Egyptian species, the Middle Eastern Dormouse, is one of the more terrestrial species, but even so, is an adept climber.

Dormice are renowned for long winter hibernation that may be up to seven months in northern Europe. It is not known if the Middle Eastern Dormouse hibernates in Egypt, but it is unlikely as its food is available year round. Dormice are omnivores, eating a wide range of food from fruit and seeds to insects, earthworms, etc., and even eggs, and small vertebrates. Another characteristic of the dormice is their wide range of vocalizations, generally related to behavior. Calls include grunts, growls, whistles, and snores.

The Middle Eastern Dormouse can be distinguished from all other Egyptian rodents by the combination of a black facial mask and long tail terminating in a brush, black above, white below, often with a white terminal tip.

MIDDLE EASTERN DORMOUSE (ASIAN GARDEN DORMOUSE) *Eliomys melanurus* Wagner, 1840
Pl. 22

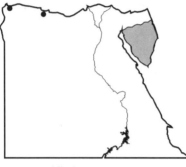

Middle Eastern Dormouse
(Eliomys melanurus)

Subspecies occurring in Egypt: *E. m. melanurus* and *E. m. cyrenaicus*.
Arabic: *Fa'r al-shagar*
Identification: Length 208–262mm; Tail 104–127mm; Weight 38.4–63g. Distinctive full-tailed rodent with diagnostic black face mask. Upper parts rusty-brown to gray along flanks. Underparts clearly delineated and whitish. Inside of legs whitish. Outside of forelegs darker than flanks. Feet white. Head boldly marked. Crown orange-brown extending down snout. Black mask through large eyes extending partly down side of snout and back to the base of the ear. Mask to underside of head and throat white. Ears rather large, sparsely haired, and with white patch above black of mask at opening. Tail long, almost as long as head and body, with first-quarter short-haired and grayish brown, and the rest of the tail bushy and black. Tip white, at least in some individuals, especially in Sinai, and underside white to grayish. Vocal with wide range of whistles and growls, and a soft call likened to snoring.

Range and status: Turkey south through Syria, Lebanon, and Israel to Arabia. Also found across North Africa from Morocco to Egypt including Sinai. In Egypt, *E. m. melanurus* found throughout Sinai except perhaps in the north. Recorded from St. Katherine, Wadi Feiran, al-Tor north to Quseima, and Nakhl. *E. m. cyrenaicus* recorded from the north coast west of Mersa Matruh.

Habitat: Mountains, cliffs, including coastal cliffs along the Mediterranean coast, stone buildings, walls, and gardens. Despite its name, can live in barren areas. Found to at least 1,700m in Sinai.

Habits: Little known. Largely nocturnal but may be active in early morning. Elsewhere, uses an old bird's nest or squirrel drey as a den or builds its own nest. Uncertain in Egypt, but in Libya has been recorded from the midst of tamarisk clumps or from the base of palms, also from settlements. Diet probably omnivorous including seeds, fruit, and insects. Breeding in Egypt little known but probably confined to spring and summer.

Associated Species: In Sinai, found in same habitat as the Cairo Spiny Mouse and Golden Spiny Mouse, Bushy Tailed Jird, and Wagner's Gerbil.

Middle Eastern Dormouse
(Eliomys melanurus)

Notes: In the past, the Middle Eastern Dormouse has been classified as a subspecies of the Garden Dormouse *Eliomys quercinus* found over much of Europe. However, it is now deemed by most authors to be distinct enough to merit full species status. For a fuller discussion on this, see Harrison and Bates (1991).

Similar species: The only other Egyptian rodent with a bushy black tail is the Bushy Tailed Jird. This can be distinguished from the Middle Eastern Dormouse by its more prominent white tip to the tail and by the total lack of a facial mask. Confusion need only arise in Sinai as it is absent from the Western Desert.

The Old World Porcupines—Family Hystricidae
11 species with 1, possibly 2, in Egypt.

The old world porcupines, and more especially the five species of crested porcupine of the genus *Hystrix*, to which both of Egypt's species belong, are amongst the largest and most distinctive rodents. No other group has taken the development of body spines to the lengths of the porcupines. The great crest of black-and-white spines along the back include black-and-white quills up to 50cm long. The flanks too are covered in spines and the tail has specially adapted hollow-tipped spines that can be rattled. Furthermore, the longer quills can be shed if attacked, leaving the assailant with a face full of spines, which, though not poisonous, can inflict a serious wound that may become infected.

Their armory of spines apart, porcupines are large, blunt-headed, short-limbed rodents. While they bear a very superficial resemblance to the much shorter-spined hedgehogs, they are entirely unrelated to these insectivores and are thought to be closest to such South American groups as the agoutis (Dasyproctidae), chinchillas (Chinchillidae), and the cavies (Caviidae). Their closest old world relatives are probably the cane rats (Thryonomyidae) of Sub-saharan Africa.

The two Egyptian species (though the Indian Crested Porcupine *Hystrix indica* has yet to be proven in Egypt) are extremely similar but widely separated geographically, which should eliminate any source of confusion.

CRESTED PORCUPINE (NORTH AFRICAN CRESTED PORCUPINE)
Hystrix cristata Linnaeus, 1758
Pl. 18

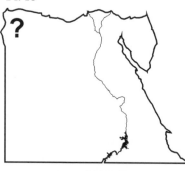

Crested Porcupine
(*Hystrix cristata*)

Probably monotypic.
Arabic: *Shayham, Duldul, Nis*
Identification: Length 770–1,020mm; Tail 120–170mm; Weight 15–27kg. Large and very distinctive rodent. Unmistakable. Thickset with coarse blackish brown fur with thick bristles on neck, shoulders, and legs, thinner ones on head. Along center of head along neck to back, a mane of long, brownish white hairs, paler toward tip, that can be erected to form a crest. From middle of back, along hind flanks and upper side of tail, there is a crest of black-and-white quills, reaching up to 40cm in length. Quills at end of tail white, short, and hollow. Head large and blunt, with small eyes and ears. Head yellow-gray in color. Legs stout with large pads and powerful claws. Voice confined to grunts, but can also rattle the tail quills.

Range and status: In a band across central Africa from Senegal to Tanzania north to Sudan in the east. In the west, found along west coast skirting the Sahara and across North Africa to Egypt. In Egypt, restricted to the area around Sallum where possibly extinct.

Habitat: In Africa, occurs in a wide range of habitat from rainforest to semi-desert. In Egypt, recorded only from cliffs north of Sallum.

Habits: Unknown in Egypt. Elsewhere, largely nocturnal but may be active at dusk. Spends day in burrow either self excavated or commandeered. Also lies up under rocks, in caves, or areas of thick vegetation. Generally singly or in pairs. Wanders along often well-defined tracks. Diet is vegetarian including roots and tubers. Will also take bones back to the burrow to gnaw, presumably for the calcium, and these may be found outside the entrance. Predators in Egypt have been exterminated (i.e., the Lion *Panthera leo*, and the Leopard *Panthera pardus*) so main threats have been humans and habitat destruction. When threatened, turns its back to threat and raises the long quills along its back, stamps its hind feet, and rattles the tail quills. Mating made possible by female flattening her quills against her back. Gestation 7–8 weeks. Litter 1–4 born in grass-lined nest within burrow. Longevity up to 20 years.

Notes: Reports of porcupines from northeastern Sinai probably relate to the Indian Crested Porcupine *Hystrix indica* found just across the border in Israel and through much of the Middle East and Arabia east to Nepal, India, and Sri Lanka—see Harrison and Bates (1991). Two individuals in the zoo at al-Arish were reportedly captured in Wadi al-Arish and were tentatively identified as this species on the basis of their dark, rather than yellow-gray, head and shoulders. Unconfirmed records of this species in the area are recorded in Osborn and Helmy (1980) and there is anecdotal evidence from local people to support the presence of porcupines.

Similar species: Porcupines cannot be confused with any other animal. Hedgehogs have much shorter spines and are far smaller. Sightings in northeastern Sinai, however, should be as carefully recorded as possible in order to confirm if the population that is possibly there relates to the Indian Crested Porcupine (which is found across the border in Israel), or, far less likely, is an isolated population of the Crested Porcupine.

The Blind Molerats—Family Spalacidae

8 species worldwide with 1 species in Egypt.

The blind molerats form a very distinct family, unmistakable in the region, but are superficially similar to the African molerats (Bathyergidae). No other group of mammals, with the possible exception of the unrelated golden moles (Chrysochloridae), shows such extreme adaptation to life underground. Blind molerats have vestigial external ears, no external tail, and eyes that are buried beneath the skin and are functionless. The lips actually seal behind the powerful incisors that it uses for digging so that it can excavate without consuming soil. The barrel-shaped body, short legs, and velvet fur (that lies neither forward nor backward) are all adaptations to life in a burrow.

As might be expected, the burrows are long and elaborate, and, according to one author, 'palatial.' They may be up to 350m long and have numerous side tunnels and storage chambers, sleeping quarters, and nursery. The mounds thrown up by the blind molerats are all that is likely to be seen of them as they rarely come above ground. The single Egyptian representative is unlikely to be mistaken for any other Egyptian mammal.

LESSER MOLERAT (EHRENBERG'S MOLERAT, MOLERAT) *Spalax leucodon* Nordmann, 1840

Pl. 27

Subspecies occurring in Egypt: *S. l. ehrenbergi.*

Arabic: *Khuld, Abu 'amaya*

Identification: Length 155–204mm; No external tail; Weight 107.2–120.4g. Very distinctive rodent, highly adapted to life underground and the only Egyptian mammal without apparent eyes. Male generally slightly larger than female. Cylindrical in form, pale brown to brown above sometimes tinged rust. Grayer below. Fur dense and soft and, like a mole, erect rather than lying forward or backward— an adaptation to burrow life. Limbs very short. Feet silvery-gray. Head broad and heavy, with broad snout, and naked muzzle. There may be a whitish band along the top of the snout. Whiskers virtually absent but a band of stiffened hairs run from each side of the muzzle toward the ear. Ears very small. Eyes completely hidden beneath skin. Tail not visible externally.

Range and status: In North Africa from Libya east. In Middle East, Israel, and Jordan north to Lebanon, Syria, Turkey, and on to southeastern Europe. In Egypt, restricted to the north coast from Alexandria west.

Habitat: Sandy areas with vegetation along the north coast, e.g., barley fields, gardens, etc. Avoids rocky areas and does not extend into the desert proper.

Habits: Almost entirely fossorial and best located by the distinctive mounds of fresh earth it throws up, up to 20cm high but much larger during breeding season. Tunnels can be very long, up to 40m in Egypt but to 350m

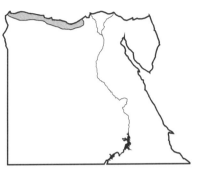

Lesser Molerat
(Spalax leucodon)

recorded elsewhere, and elaborate with chambers for sleeping, storage, and defecating. Burrow system described by one author as 'palatial.' Uses its powerful incisors to dig rather than its forelimbs. Seems to dig more during wetter seasons, less in the summer when it tends to dig deeper. Shows a preference for harder soils. Diet is vegetarian feeding largely on roots, bulbs, and tubers. Will hoard food in storage chambers. Underground habits protect it from most predators but when it does come to the surface, generally at night to forage or find a mate, etc., it is vulnerable to birds of prey and owls. Defends itself by biting or burrowing rapidly. Gestation 28+ days. Litter size 3–4 but up to 9 elsewhere in region with 1 litter per year. Excavates a special breeding chamber throwing up a mound of earth much greater than from normal excavations. Activity not inhibited by human presence.

Associated Species: Found in the same habitat as the Anderson's Gerbil, Shaw's Jird, Lesser and Greater Egyptian Jerboas, Four-toed Jerboa, and Fat Sand Rat.

Notes: Osborn and Helmy (1980) refer North African and Middle Eastern molerats to Ehrenberg's Molerat *Spalax ehrenbergi*. As variation seems continuous over the current species' range, Harrison and Bates (1991) has been followed, treating the Egyptian Lesser Molerat as a subspecies of *Spalax leucodon*.

Similar species: No other Egyptian mammal lacks external eyes or is so completely adapted to life underground. There are no moles in Egypt. Unmistakable.

The Lagomorphs—Order Lagomorpha

There are 58 species of Lagomorphs worldwide of which 1 occurs in Egypt.

The Lagomorpha, better known as the rabbits, hares, and pikas, is an order of mammals superficially similar to the rodents but which differs primarily on cranial characteristics; in particular, lagomorphs have four incisors in the upper jaw (although only two are functional) as opposed to two in the rodents. The order is divided into two families, the pikas (Ochotonidae), relatively short-eared lagomorphs found in the mountainous regions of Asia and North America, which are not represented in Egypt, and the much more familiar rabbits and hares (Leporidae) of which one species occurs wild in Egypt, though the rabbit is familiar as a domestic animal.

The Rabbits and Hares—Family Leporidae

44 species worldwide with 1 in Egypt.

Most rabbits and hares, including the Egyptian representative, are instantly recognizable by their greatly elongated and mobile ears, large eyes set high on the head, nostrils in the form of closeable slits, large and powerful hind feet, very short tail, and dense, soft fur. All escape predators by using their greatly developed hind limbs and feet to run at speeds of up to 80kmph often zigzagging to further confuse the attacker. The front legs are shorter and the front feet bear five toes as opposed to the hind feet with four. The undersides of the feet are covered with hair, in contrast to most rodents, which probably gives the rabbit or hare more purchase on soft ground and cushions the impact on hard ground.

The most distinctive feature of the family, however, must be the ears. These are generally long and large and it is not surprising that their sense of hearing is acute. However, the ears also have another use, that of temperature control. The hares can control the flow of blood through the enormous ears so as to regulate temperature loss or gain. The eyes, too, are large, and set high on the head, giving them a wide field of vision. Not surprisingly for animals with large eyes and ears, the rabbits and hares are predominantly nocturnal, though, where undisturbed, they may be active by day. For further information, see the species description below. The Cape Hare *Lepus capensis* is the only species represented wild in Egypt. References to rabbits will be of domestic animals in agricultural areas and of either hares or hyraxes (see page 141)—that are superficially rabbit-like—in mountainous areas. The Cape Hare is divided into a great many subspecies over its huge range, of which four occur in Egypt. It is difficult to confuse with any other Egyptian mammal.

CAPE HARE (BROWN HARE, ARABIAN HARE) *Lepus capensis* Linnaeus, 1758
Pl. 18

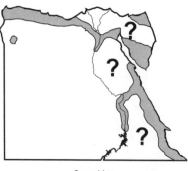

Cape Hare
(*Lepus capensis*)

Subspecies occurring in Egypt: *L. c. sinaiticus, L. c. aegyptius, L. c. isabellinus,* and *L. c. rothschildi.*

Arabic: *Arnab al-kab*

Identification: Length 349–567mm; Tail 49–102mm; Weight: no data from Egypt. Up to 7kg elsewhere, but generally much lighter. A large, rodent-like mammal with unmistakable long, erect ears, large hind feet, and short tail. Yellow-brown to buff-gray above, grayer in Sinai, paler along flanks and white below. Coat dense and soft. Legs and feet as upper parts. Inside leg white. Hind feet greatly enlarged, generally over 10cm long (note tracks). Sole with long hair. Head with blunt snout. Throat buff. Eyes prominent and surrounded by whitish eye ring. Ears very large and elongated, pale brown sometimes tipped with black. Inside paler. Tail short, blackish above, white below, held erect when running exposing white underside to the observer. When alert, sits upright with ears erect. Powerful runner.

Range and status: Huge range over much of Europe and Russia east to China. Also Arabia and Middle East, North Africa and Africa south of the Sahara, except West Africa and rainforest regions of central Africa. Introduced to the United States, Australia, New Zealand, etc. In Egypt, subspecies largely separated geographically. *L. c. sinaiticus* restricted to Sinai including St. Katherine, Wadi Feiran, along west coast and north to Quseima and al-Arish east to Rafah. Found in Zaranik Protected Area. *L. c. capensis* in Eastern Desert north of Marsa Alam to Cairo–Suez Road, including Wadi Digla Protected Area and Wadi Rishrash. *L. c. isabellinus* from southern Eastern Desert including Gebel Elba region, Wadi Allaqi, and possibly shores of Lake Nasser. *L. c. rothschildi* in Western Desert from along entire north coast to Delta margin, Wadi Natrun, Siwa, and the Fayoum (status unknown) south to Aswan where it may occur with *L. c. isabellinus*. Distinguished by slightly smaller size, more reddish color, and lack of black-tipped ear. Probably even more widespread than records suggest, though hunted and also killed on roads.

Habitat: Open country of all types including open desert as long as there is sufficient vegetation for food. Open agricultural areas such as barley fields, wadi floors, and even mountainous areas in Eastern Desert and South Sinai. In Siwa, recorded from clumps of vegetation at desert margin.

Habits: Probably largely nocturnal but also active by day even in vicinity of human habitation. Does not excavate a burrow (though elsewhere in region may use the burrow of another species) but lies up in a shallow depression known as a form. Here, with ears held flat against the back and protective coloring, very difficult to see. Vegetarian diet, very varied. Predators probably include the Striped Polecat *Poecilictis libyca*, Wild Cat *Felis silvestris*, Cheetah *Acinonyx jubatus* (in Qattara Depression), large birds of prey, and the Eagle Owl *Bubo bubo*. Also a victim of road kills. Normal gait when feeding is a relaxed lope. When threatened or disturbed, runs for safety jinking in zigzags and with tail raised. Hearing excellent, sight and smell good. Generally solitary, holding a home range. May deposit fecal pellets in regularly used latrines. Gestation 42 days. Breeding information scant from Egypt with records from spring, but may have more than 1 litter of 1–4 young (leverets) a year.

Notes: There are no wild rabbits in Egypt, though both the Cape Hare and the Rock Hyrax *Procaria capensis* have been referred to as rabbits.

Similar species: Difficult to confuse with any other mammal in the country. The Four-toed Jerboa *Allactaga tetradactyla* has long ears but is much smaller and has a long, slender tail. Hyraxes have short ears and are confined to rocky areas.

Order of Plates

Plate 1

The Hedgehogs

Large insectivores with distinctively spiny coats. Unlikely to be mistaken for any other group of Egyptian mammals. Spiny mice *(Acomys spp.)* are much smaller with much less prominent and extensive spines and the porcupines *(Hystrix spp.)* are far larger with much longer and bolder black-and-white spines. *Hemiechinus* sp. has one representative in Egypt and it is the most common hedgehog. It is distinguished from the *Paraechinus* sp. by smaller size, longer ears, and lack of 'parting' of spines on crown.

1. Long-eared Hedgehog   p. 26
Hemiechinus auritus
Small, long-eared, long-limbed hedgehog. Agile. Without distinctive patterning around face. Only other hedgehog within range is the Desert Hedgehog *Paraechinus aethiopicus,* see differences above.

The taxonomy of the *Paraechinu* genus is open to much debate. Fo differentiation from *Hemiechinus,* see above. The three hedgehogs described below are treated as subspecies o *Paraechinus aethiopicus,* but as far a field identification is concerned, al are geographically isolated, preclud ing confusion. The facial pattern is a useful identification pointer but i also variable.

2. South Sinai Hedgehog   p. 28
Paraechinus aethiopicus dorsalis
The only hedgehog in South Sinai Unmistakable.

3. Desert Hedgehog   p. 28
Paraechinus aethiopicus deserti
For differences with the Long-eare Hedgehog, see above.

4. Ethiopian Hedgehog   p. 2
Paraechinus aethiopicus aethiopicus
The only hedgehog recorded from th southeast of Egypt. Unmistakable.

1

——————— Plate 2 ———————

The Shrews

Small to extremely small insectivores. Superficially similar to the unrelated mice but readily distinguished by the long, pointed snout, tiny eyes, and smaller ears. Very active mammals foraging at all times of the day and night due to their voracious appetite. Most with very few records but possibly underrecorded. Note size, coloring, and length and extent of bristling on the tail.

1. Greater Musk Shrew   p. 30
Crocidura flavescens
Large shrew and the only shrew at all widespread in Egypt. Large size distinguishes it from all other shrews in Egypt except the very rare House Shrew *Suncus murinus*. Overall dark, grayish below. Tail dark with bristles extending along the first two-thirds of its length.

2. Dwarf Shrew   p. 32
Crocidura nana
Small shrew. Rather nondescript with grayish rather than whitish underparts. Tail dark and long, c. 60% of head and body length. Tail with abundant bristles.

3. Lesser White-toothed Shrew p. 33
Crocidura suaveolens
Very similar to Dwarf Shrew but note paler underparts and feet (whitish) and proportionately shorter tail, c. 50% of head and body length. Bristles on tail extend along entire length. Teeth lack reddish tip.

4. Flower's Shrew   p. 32
Crocidura floweri
Small shrew tinged brownish above, rather than grayish. Whitish below. Tail c. 75% of head and body length. Bristles extend along basal half of tail. Very rare.

5. House Shrew   p. 35
Suncus murinus
Large and robust shrew. Large size distinguishes it from all other shrews except Greater Musk Shrew. Head massive. Tail roughly half of head and body length with thick stock tapering to tip. Probably introduced and commensal.

6. Savi's Pygmy Shrew   p. 36
Suncus etruscus
Minute shrew (the smallest terrestrial mammal). Brown above, whitish below. Tail slender and proportionately long, c. 75% of head and body length, with bristles along entire length. Ears large and stand well clear of the fur.

⑥

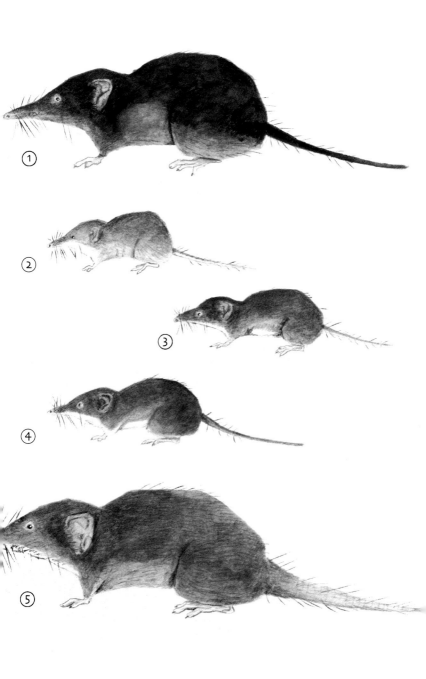

Plate 3

The Egyptian Fruit Bat
and the Free-tailed Bats

The fruit bats, suborder Megachiroptera, are represented by a single, highly distinctive representative in Egypt.

1. Egyptian Fruit Bat   p. 38
Rousettus egyptiacus
Largest Egyptian bat. Size alone should make the Egyptian Fruit Bat unmistakable. Very short tail and tail membrane much reduced. Face fox-like with large eyes. In the hand, the second finger is clawed, unique amongst Egyptian bats. Ears simple. Tragus absent.

————

The free-tailed bats *Tadarida* spp. are represented in Egypt by two similar species best differentiated by size and dental characteristics. Behavior is distinctive. Both species are agile and adept on the ground, often scuttling into crevices rather than flying.

2. European Free-tailed Bat   p. 66
Tadarida teniotis
Large, high-flying, slender-winged bat. Largest Egyptian insectivorous bat. Wings very different from previous species, being long and slender. Ears very large and forward pointing, do not meet at base. Tragus very broad and unnotched. Snout naked. Very dark above, paler below. Tail only enclosed in flight membrane toward the base.

3. Egyptian Free-tailed Bat   p. 68
Tadarida aegyptiaca
Large (though smaller than previous species) high-flying, slender-winged bat. Very similar to, but smaller than, the previous species. Ears large and forward pointing, do not meet at base. Tragus not as broad as in previous species and notched. Snout naked. Paler above than previous species. Tail similar but generally proportionately shorter.

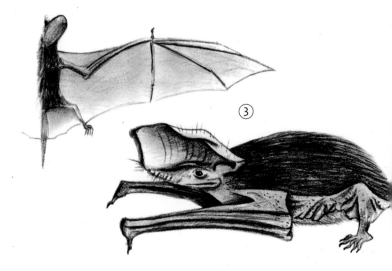

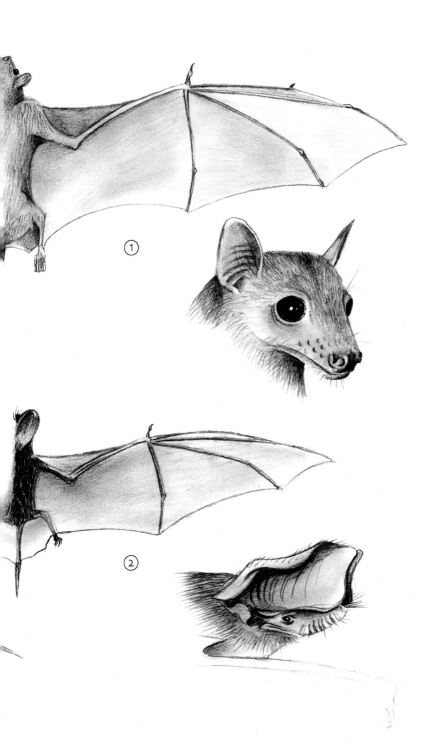

①

②

Plate 4

The Rat-tailed Bats, the Sheath-tailed Bats, and the Egyptian Slit-faced Bat

The rat-tailed bats of the family Rhinopomatidae are often regarded as the most primitive of the insect-eating bats. Their slender wings are characterized by relatively very short fingers. The snout is simple and rather pig-like. The ears have a well-developed tragus and are connected by a membrane across the forehead. Their most distinctive feature is the long, very slender tail extending well beyond the flight membrane.

1. Larger Rat-tailed Bat   p. 41
Rhinopoma microphyllum
Slender-limbed bat with long, very slender tail. In flight, indeed also at rest, tail so slender as to be barely visible, but also note much reduced interfemoral membrane. Generally clings to wall surface, cave walls, etc., with head down and tail curled over body. Eyes prominent. Forearm length generally greater than tail length.

2. Lesser Rat-tailed Bat   p. 41
Rhinopoma hardwickii
Can only be mistaken for the previous species (which it closely resembles), but is smaller and more lightly built. Key distinction in hand is that the tail length is greater than forearm length. Much more common than previous species.

The sheath-tailed and tomb bats of the family Emballonuridae are small- to medium-sized bats with large eyes, rather long, slender ears, and a simple muzzle unembellished by noseleaves. Tragus distinctive. Diagnostic feature is that the tail is embedded in the interfemoral membrane for some of its length and then emerges, but does not extend, in its 'free' state beyond the membrane.

3. Geoffroy's Tomb Bat   p. 43
Taphozous perforatus
Medium-sized bat with unembellished, naked facial region and no gular sac. Brownish above, grayer below with long, narrow ears and distinctive tragus. Hair extends all over belly and back, and onto interfemoral membrane.

4. Egyptian Sheath-tailed Bat p. 44
Taphozous nudiventris
Larger than previous species. Facial area naked, muzzle simple. Small gular sac. Ears long and narrow with distinctive tragus. Brownish above, paler and grayer below. Lower belly and back are absolutely naked, a characteristic unique amongst Egyptian bats.

The slit-faced bats of the family Nycteridae are represented in Egypt by one very distinctive species. Although the large ears give them a superficial resemblance to certain vesper bats, e.g., Gray Long-eared Bat *Plecotus austriacus*, the slit along the muzzle and T-shaped terminal tail vertebra are unique amongst Egyptian bats.

5. Egyptian Slit-faced Bat   p. 45
Nycteris thebaica
A large-eared and broad-winged bat. Features cited under family description render this bat unmistakable. Lappets visible in live specimens. For distinctive flight pattern, see text.

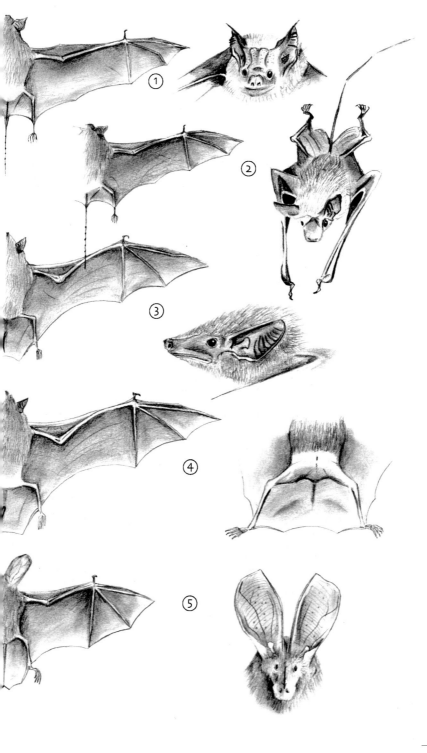

Plate 5

The Horseshoe Bats
and the Trident Leaf-nosed Bat

Medium- to small-sized bats distinguished by an elaborate noseleaf related to echolocation. It is the detailed structure of this noseleaf that is important in distinguishing between the four species, themselves divided into two families, the horseshoe bats (Rhinolophidae) and the leaf-nosed bats (Hipposideridae), based, at least in part, on the structure of this noseleaf.

The horseshoe bats (Rhinolophidae) in Egypt all belong to the genus *Rhinolophus*. They are readily distinguished from all other Egyptian bats in the hand by the noseleaf with its diagnostic 'horseshoe' at the base. The complex structure of this noseleaf is key to identifying the species. Told from other bats by the presence of this leaf, absence of the tragus, distinctive ear shape, entirely enclosed tail (within the tail membrane) and, at roost, hanging stance (very weak forelimbs render most species almost helpless on the ground). In flight, broad winged.

1. Arabian Horseshoe Bat   p. 48
Rhinolophus clivosus
A large, widespread (but apparently scarce) horseshoe bat. Dark, with dark flight membranes. Underside dull gray. See diagram for noseleaf.

2. Lesser Horseshoe Bat   p. 49
Rhinolophus hipposideros
Small horseshoe bat with only one record

from Egypt (St. Katherine area). Delicate build and small size distinctive, but poor field characteristics. Color variable. See diagram for noseleaf and shape of sella, which are unique amongst Egypt's *Rhinolophus* spp.

3. Mehely's Horseshoe Bat   p. 50
Rhinolophus mehelyi
Medium-sized horseshoe bat with few records from Egypt. Sella and lancet distinct. Paler than Arabian Horseshoe Bat. Identification must rely on detailed analysis of the noseleaf. Real confusion with the extralimital (but potential) Mediterranean Horseshoe Bat *Rhinolophus euryrale*. Again, noseleaf structure is key. See diagram.

The leaf-nosed bats (Hipposideridae), similar to the horseshoe bats, lack a tragus but differ in the structure of the noseleaf. There is only one confirmed Egyptian representative.

4. Trident Leaf-nosed Bat   p. 52
Asellia tridens
Medium-sized leaf-nosed bat with very different noseleaf structure than true horseshoe bats. In the field, a very pale bat. Tail extends further beyond tail membrane than in true horseshoe bats. Diagnostic feature is the structure of the noseleaf with no 'horseshoe' structure but a distinct 'blunt, sharp, blunt' triad of structures above the noseleaf. Based on records, more widespread and common than any horseshoe bat.

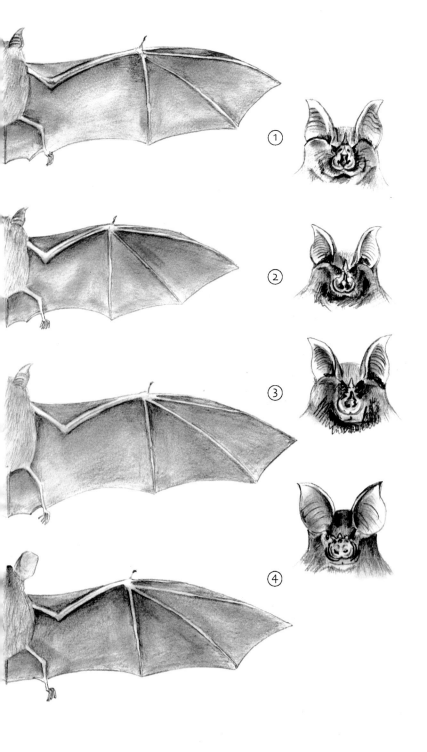

①

②

③

④

Plate 6

The Vesper Bats (1)

The vesper bats of the family Vespertilionidae form the largest bat family. While they differ superficially in a variety of ways, particularly in the size, shape, and complexity of the ears, they share certain characteristics. They are all small- to medium-sized, rather mouse-like, with small eyes, ears normally separate on the forehead and with a distinct tragus, and a simple muzzle. There is no slit or noseleaf. The tail is entirely enclosed within the interfemoral membrane or extends beyond it only very slightly.

Five of Egypt's vesper bats belong to the genus *Pipistrellus* and include some very rare (or rarely encountered) species. While dental and cranial characteristics are probably necessary for positive identification of the more similar pipistrelles, others have distinctive features or can be separated by range though, with so little known about Egypt's pipistrelles, their true ranges are probably more extensive than currently known and species previously considered geographically separated may overlap. In the hand, the tragus structure is important.

1. Kuhl's Pipistrelle   p. 55
Pipistrellus kuhlii
Largest Egyptian pipistrelle and by far the most widespread. Very variable in color but, in most specimens, the trailing edge of the flight membrane from the foot to the tip of the fifth digit is pale. For tragus, see figure.

2. Desert Pipistrelle   p. 56
Pipistrellus aegyptius
Slightly built pipistrelle but with some size overlap with previous species. Pale brown above, paler below. So far only known from Luxor south. For tragus, see figure.

3. Pygmy Pipistrelle   p. 57
Pipistrellus ariel
Tiny pipistrelle known from very few specimens. Wing membranes pale brown (see following sp.). So far only known south of Aswan. For tragus, see figure.

4. Bodenheimer's Pipistrelle   p. 58
Pipistrellus bodenheimeri
Another tiny pipistrelle with uniformly dark wing membranes. Dark wing membrane makes this species easier to distinguish than other Egyptian pipistrelles. This contrasts with the paler tail membrane. Otherwise very pale, white, tinged buff. Long-haired. For tragus, see figure.

5. Rüppell's Bat   p. 59
Pipistrellus rueppelli
Only slightly smaller than Kuhl's Pipistrelle. Wing membranes almost a reverse of much smaller previous species, being pale gray with darker limbs. Body notably bicolored, gray-brown above and pure white below. Ears and muzzle blackish. For tragus, see figure.

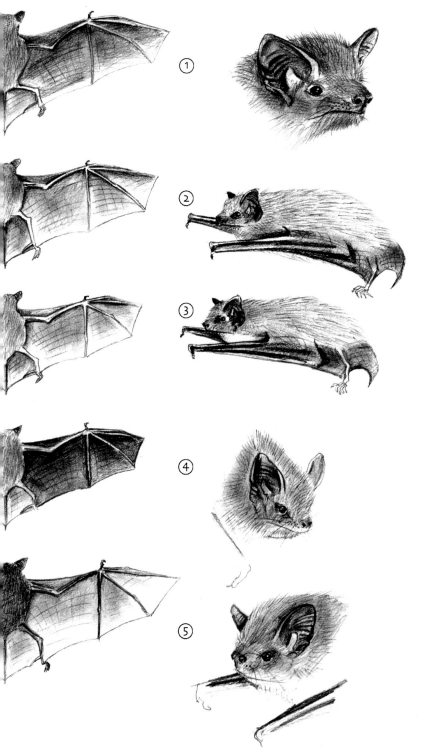

_____ Plate 7 _____

The Vesper Bats (2)

1. Botta's Serotine Bat   p. 60
Eptesicus bottae
Largest, small-eared vesper bat in Egypt. No noseleaf and simple ears. Larger than any *Pipistrellus* spp. and with tail extending further beyond flight membrane. Tragus short and narrow.

2. Hemprich's Long-eared Bat p. 61
Otonycteris hemprichii
Large bat with huge ears (up to one-third of total length). Can only be mistaken for the Gray Long-eared Bat *Plecotus austriacus*, but distinguished by much larger size and ears that do not meet on the forehead. For habitat difference, see text.

3. Schlieffen's Bat   p. 63
Nycticeinops schlieffeni
Very small bat, very similar to the *Pipistrellus* spp. Size alone distinguishes it from all other species apart from *Pipistrellus*. From these, note short and narrow tragus. Also, tail entirely within flight membrane, in most *Pipistrellus* spp. last vertebra extends beyond membrane.

4. Arabian Barbastelle   p. 64
Barbastella leucomelas
Small- to medium-sized bat with very distinctive ear and nose structure. Ears shorter than the two long-eared bat species and joined at the forehead. Tragus relatively large and well haired. Nose without leaf and very flattened and compact.

5. Gray Long-eared Bat   p. 65
Plecotus austriacus
Small bat with huge ears. Ears only slightly shorter than forearm. Can only be mistaken for Hemprich's Long-eared Bat, but much smaller and ears meet at base of forehead. Tragus with no lobes or notches. Only very tip of tail emerges from the flight membrane.

① ② ③ ④

Plate 8

The Wolf and the Jackal

Large canids that can be distinguished from the foxes by size, the relatively shorter and much less full tail, and the relatively smaller ears. In Egypt, the Wolf *Canis lupus* and the Jackal *C. aureus* are extremely similar and there is still debate as to the status of the former in the country. The usual differentiation by size is open to discussion and no single cranial characteristic holds true (for discussion, see text). According to one expert, the best distinction in the field is gait: Wolves lope while Jackals trot.

1. Wolf   p. 72
Canis lupus

Large canid, though Wolves from the region are much smaller and lighter in build than those from, e.g., northern Europe. Lanky, long-limbed. Predominantly grizzled gray with buff, especially on shoulder and front of thigh. Tail grayish with dark tip. So far, recorded from Sinai only.

2. Jackal   p. 70
Canis aureus

Largish canid, though overlaps in size with previous species. Shaggy, dog-like, grizzled gray-brown with longer erectile hairs along back. Paler below. Legs buffish. Tail, as dorsum, darker along the top and with dark tip. Only one confirmed record from northeast Sinai.

Plate 9

The Foxes

Small- to medium-sized canids readily distinguished from the Wolf *Canis lupus* and Jackal *Canis aureus* by smaller size and lighter build, bushier, fuller tail or brush, more pointed snout, and proportionately larger ears. Key features include size, color of back of ear, facial pattern, and tail color and proportion. 'Violet' gland on dorsal side of tail can be prominent.

1. Red Fox   p. 74
Vulpes vulpes
Largest fox in Egypt. Variable. In winter, appears quite dark (as in plate), in summer, paler. Ruddy to gray-brown above, variably tinged reddish. Throat and belly dark. Ears large, though proportionately smaller than following species. Back of ears blackish. Tail full and bushy, white tipped.

2. Rüppell's Sand Fox   p. 76
Vulpes rueppelli
Slim, lightly built fox. Rather pale. Reddish along back with buffish flanks and white underparts. No black on fore- or hind limbs. Ears proportionately very large.

Back of ears buffish. Distinct facial pattern more strongly marked than in previous species, rufous around eyes, and dark patch to the base of slender snout. Tail very full, buffish flecked black. White tip.

3. Blanford's Fox   p. 77
Vulpes cana
Small, richly furred fox with very large tail. Dark from base of neck, along back and upper side of tail. Flanks pale gray, russet below. Ears proportionately large. Back of ears buffish. Facial pattern as in previous species, but dark patch from eye narrower. Snout very slender. Tail very full and, proportionately, the longest of the Egyptian foxes, blackish above, and generally dark tipped.

4. Fennec Fox   p. 79
Vulpes zerda
Very small, pale fox. Uniform creamy, slightly darker along back, whitish below. Ears proportionately huge, back of ear pale. Facial pattern not strongly marked. Tail bushy and full, though less so than in other foxes. Tail tip rather pointed and blackish.

②

Plate 10

The Mustelids

The mustelids, or weasel family, are slim, active carnivores represented in Egypt by four species of weasels and polecats.

1. Zorilla   p. 84
Ictonyx striatus

Large, strikingly marked weasel. Underside uniform black, dorsum white with three, regular black stripes—the center one continuing onto the tail. Head black with white patches below ears and on forehead. Southern Eastern Desert only. Can only be mistaken for geographically separate Striped Weasel *Poecilictis libyca*.

2. Striped Weasel   p. 82
Poecilictis libyca

Smallish, strikingly marked weasel. Shaggy coat, black below, and black-and-white above with stripes disjointed and irregular. Head distinctively patterned with an unbroken white band across the forehead. So far, recorded from the Delta and northern Western Desert only, so geographically separated from previous species.

3. Egyptian Weasel   p. 81
Mustela subpalmata

Very small, elongated weasel. Uniform, dull brown above, pale below with irregular demarcation. Very slender and agile. Tail relatively short and held erect when running. In Egypt, largely commensal.

4. Marbled Polecat   p. 85
Vormela peregusna

Strikingly patterned weasel. Deep chocolate-brown below, orange-buff above, complexly marked with darker brown. Head strikingly marked, dark brown with white muzzle, band across forehead and around the rim of the ears. Tail full, base brownish, paling to cream and with dark tip. Recorded from northeast Sinai only.

③

Plate 11

The Egyptian Mongoose
and the Small-spotted Genet

The Egyptian Mongoose *Herpestes ichneumon* and the small-spotted Genet *Genetta genetta* are the Egyptian representatives of the families Herpestidae and Viverridae, respectively. Though they differ greatly in appearance, both are slim, short-legged, long-tailed carnivores, superficially weasel-like in appearance.

1. Egyptian Mongoose   p. 89
Herpestes ichneumon
Weasel-like but much larger and shaggier. Confined to the Nile Delta south to Cairo and the Fayoum. Slender, shaggy-coated carnivore. Like weasel but much larger. Grizzled gray-ocher above due to dark and pale banding of hairs. Below more uniform. Snout fairly long and narrow, short haired, and blackish. Ears broad and rounded. Tail long, appears thicker at base due to longer hairs, grizzled as upper parts, and with black tip, slightly tufted.

2. Small-spotted Genet   p. 87
Genetta genetta
Between weasel and cat in appearance, agile with very long, full tail. In Egypt, only recorded from the very southernmost parts of the Eastern Desert. Slender, elongated carnivore. Coat short, gray patterned with extensive black markings. Upper parts spotted and striped. Pale below with some spotting. Face strikingly marked black and white, especially around eyes. Ears larger than in previous species. Tail long and full, distinctly banded dark and pale, tapering toward tip rather than tufted.

Plate 12

The Striped Hyena and the Aardwolf

Although superficially similar, the Striped Hyena *Hyaena hyaena* and the Aardwolf *Proteles cristatus* are unlikely to be confused in the field. Some authors assign the two to separate families, the Hyaenidae and the Protelinae, respectively.

1. Striped Hyena p. 92
Hyaena hyaena

Large, dog-like carnivore with distinctively patterned coat and sloping hindquarters. Forequarters powerful, with heavy jaws, and large ears. Muzzle blunt and black. Overall coat is shaggy, pale gray, and striped, especially along the flanks, hindquarters, and neck. Crest along the back can be raised in anger or alarm. Tail rather short and pale. Much larger and more heavily built than the Aardwolf and more strikingly patterned than the large canids.

2. Aardwolf p. 94
Proteles cristatus

Like a much smaller, slimmer, and more lightly built version of previous species. Slender-limbed and long-necked mammal with a shaggy coat, patterned very much as the above species. Also has an erectile crest along the back. Face and muzzle uniformly dark. Ears large and rather pointed. Tail full and dark tip-ped. Very few records from Egypt, all from the southeasternmost part of the country.

①

②

Plate 13

The Smaller Cats

While the Swamp Cat *Felis chaus* and the Sand Cat *F. margarita* are fairly distinctive and found in very different habitats, the Wild Cat *F. silvestris* can be very similar to tabby-type domestic/feral cats. Extreme caution is needed in identification, especially in the vicinity of human habitation. This is compounded by the fact that interbreeding with feral cats has been identified as a major threat to the true Wild Cat.

1. Swamp Cat   p. 98
Felis chaus
Long-limbed, short-tailed cat. Significantly larger than the Wild Cat or domestic cat. Coat rather uniform with faint markings, though distinct 'tear' stripe running down from eye. Yellowish brown above, paler below, faintly spotted. Ears reddish brown behind with black tips and short tufts. Tail relatively short, black tipped, with two rather narrow black bands.

2. Wild Cat   p. 96
Felis silvestris
Wild ancestor of the domestic cat to which it can be very similar. Smaller than the previous species. Buffish above, with striping on flanks and spotting on paler belly. Upper legs striped. *F. s. tristrami* from North Sinai is darker overall. Lankier than domestic cat. 'Tear' stripe from eye, pale brown, less distinct than in the previous species. Ears reddish brown behind, with no tuft. Tail relatively long, as in the domestic cat, with dark rings and black tip.

3. Sand Cat   p. 100
Felis margarita
Elusive, very small, pale cat. Smaller than domestic cat, though long coat makes it look larger than it is. Pale, yellow-buff, indistinctly marked except for dark bands on forelegs. Orange-buff 'tear' stripe. Ears relatively very large and broad-based, tip dark behind, with no tuft. Tail indistinctly striped with distinct dark tip. Soles of feet heavily furred dark brown, obscuring pads.

① ② ③

Plate 14

The Larger Cats

Recent records are very sparse for all three species of larger cats. The Cheetah *Acinonyx jubatus* is probably already extinct in Egypt with recent expeditions failing to sight it. The Leopard *Panthera pardus* is probably extinct in South Sinai but could still occur in the southern Eastern Desert. The Caracal *Felis caracal* may be more numerous than recent records suggest. Reported sightings of any of these species should be very carefully documented.

1. Cheetah   p. 104
Acinonyx jubatus
Large, slim, long-limbed cat. Confirmed only from northern Western Desert where, at best, very rare. Lanky build very different from the next species. Pale yellowish above, marked with solid round spots not arranged in rosettes. Underside white. Head relatively small and rather flat. Ears small. Very distinct black 'tear' mark. Tail long, thickening toward end, with white tip.

2. Leopard   p. 106
Panthera pardus
Powerfully built, robust, big cat. Possibly still in mountains of South Sinai and southern Eastern Desert. One old record from Western Desert. Superficially similar to previous species but, in reality, very different. Much heavier build. Buffish to yellow-buff above, white below. Heavily spotted, spots on back flanks and upper legs forming distinct rosettes. Head powerful, spotted but lacks 'tear' stripe of previous species. Ears strikingly black and white behind. Tail long, spotted as body, spots fusing to form bands toward tip. Very distinct white underside to tip of tail.

3. Caracal   p. 102
Felis caracal
Short-tailed, uniform cat with long ear tufts. Eastern Desert and Sinai with very scattered records. About size of Swamp Cat but uniform buffish to reddish brown above. Paler below, sometimes faintly marked. Head strikingly marked with dark stripe running from behind eye to muzzle. Dark patch behind muzzle. Ears large, black behind and with elegant, elongated tufts, much longer than in the Swamp Cat. Tail conspicuously short with no barring and no dark tip.

Plate 15

The Baleen Whales, the Sperm Whale, and the Blackfish

Large cetaceans, all over 4m, but differ widely between groups. See individual descriptions.

1. Fin Whale   p. 113
Balaenoptera physalus
Largest whale recorded from Egyptian waters. Huge (up to 25m) but slimline, able to swim very fast. Dark above, pale below with head pattern diagnostic. Lower lip and upper throat white on right side, dark on the left. Dorsal fin small, falcate, and set well back on body. Up to 100 throat grooves. Single head ridge.

2. Sei Whale   p. 115
Balaenoptera borealis
Large whale, up to 16m (unconfirmed up to 21m), but considerably smaller than previous species and lacking the asymmetrical head pattern. Dark above, pale below. Thick tailstock. Dorsal fin tall, erect and slender, and further forward than in previous species. Up to c. 60 throat grooves. Single head ridge (Bryde's Whale *Balaenoptera edeni* could conceivably turn up in the Red Sea and is very similar but with three head ridges).

3. Humpback Whale   p. 116
Megaptera novaeangliae
Large, dark whale with huge pectoral fins. Up to 15m long and more heavily built than previous two species though with narrower tailstock. Generally very dark above with pale below variable, some individuals being all dark. Dorsal fin small and low. Up to 36 throat grooves. Head with distinctive knobs, especially on chin. Pectoral fins very long, up to one-third of body length, white on underside, and with knobby leading edge. Tail flukes with serrated trailing edge.

4. Sperm Whale   p. 118
Physeter macrocephalus
Large whale with huge head and no dorsal fin. Large size, up to 18m (largest toothed whale). Head huge, blunt, and with single blowhole set to left side. Thick tailstock. Generally dark with curious wrinkled texture to much of skin. No throat grooves but narrow lower jaw often obscure. No dorsal fin but dorsal 'hump' and a series of smaller 'humps' to base of tail.

5. Short-finned Pilot Whale   p. 132
Globicephala macrorhynchus
Up to 6.5m long, dark with bulbous forehead. This and the next species are basically large dolphins. All dark with variable pale patch on throat and belly. Thick tailstock. Dorsal fin curved backward, rounded, and with concave trailing edge. Pectoral fins rather long and pointed.

6. False Killer Whale   p. 133
Pseudorca crassidens
Up to 6m long, dark with slender, non-bulbous head, and rounded beak. All dark interrupted only by variable pale throat patch which may be absent. General form streamlined with slender tail stock. Dorsal fin generally more erect and pointed than in previous species. Pectoral fins short and pointed.

_____ Plate 16 _____

The Dolphins

Small- to medium-sized toothed whales (for larger blackfish, e.g., the Orca *Orcinus orca*, see plate 17). All species recorded in Egyptian waters have a more or less distinctive beak with the exception of Risso's Dolphin *Grampus griseus*. Important features include size and shape of dorsal and pectoral fins, body patterning and color, and behavioral characteristics, for which see text.

1. Common Dolphin   p. 123
Delphinus delphis

Small- to medium-sized, slim dolphin, very variable. Averages around 2m. Distinctive 'hourglass' pattern on flanks with narrow point of hourglass below dorsal fin, often forepart of hourglass described as tan, but this seems faint in Red Sea specimens so color is not reliable. No spotting. Dorsal fin prominent with trailing edge concave. Flippers dark, slightly pointed. Beak slender, often white tipped.

2. Indo-Pacific
Humpback Dolphin   p. 126
Sousa chinensis

Medium, stocky dolphin. Up to 2.8m. Obscure to obvious hump on back crowned by small, rather rounded dorsal fin. Color highly variable but unspotted and generally darker above and on flanks. No distinct facial pattern. No pattern on flanks. Beak slender, mouth line straight. Flippers broad and blunt, relatively short.

3. Spinner Dolphin   p. 128
Stenella longirostris

Small, slim, very active dolphin. Averages less than 2m. Darkest above, paler gray flanks and paler below, but may be uniform. Unspotted. Dorsal fin angular and erect, not backswept. Beak long and slender, black tipped. Forehead lacks distinct break above beak. Long, pointed flippers. Behavior distinctive.

4. Pantropical Spotted Dolphin p. 127
Stenella attenuata

Small- to medium-sized, distinctively patterned dolphin. Averages around 2m. Most similar to Common Dolphin but lacks hourglass pattern. Tricolored with dark cape, gray flanks, and pale underside. Flippers dark. Dorsal fin distinct with concave trailing edge. Mature animals generally spotted, sometimes heavily so. Beak slender with white tip.

5. Bottle-nosed Dolphin   p. 124
Tursiops truncates

Large, rather stocky dolphin. Up to nearly 4m, though average less. No distinct patterning, darker above, paler below with dark cape. Not spotted except on underside in some older dolphins. Beak shorter than in previous species and heavier, but still distinct. Dorsal fin falcate and prominent.

6. Risso's Dolphin   p. 130
Grampus griseus

Large dolphin with no beak. Up to nearly 4m. Unique amongst Egyptian dolphins in that it has no beak (though see much darker blackfish). Dull gray above, pale below often showing distinct heavy scarring. Head blunt with no beak and, especially in older animals, very pale to white. Dorsal fin with concave trailing edge prominent. Slender tail stock. Flippers long, slender, and pointed.

Plate 17

The Dugong
and the Southern Red Sea Cetaceans

The Dugong *Dugong dugon* is the only Egyptian representative of the Sirenia, which includes the extinct Steller's Sea Cow *Hydrodamalis gigas* and the manatees of the family Trichechidae. They are unrelated to the cetaceans and can be distinguished because they lack a dorsal fin. The remaining cetacean species, namely, the Orca (Killer Whale) *Orcinus orca*, the Rough-toothed Dolphin *Steno bredanensis*, and the Striped Dolphin *Stenella coeruleoalba*, have yet to be confirmed in the Egyptian Red Sea but have been recorded further south. Given the rapid increase in the number of dive operations based in and visiting Egypt's southern Red Sea, there is a real chance that these species could be recorded.

1. Dugong   p. 135
Dugong dugon

Bulky, slow-moving sea mammal. Grazes on beds of sea grass. Up to 4m. Robust with blunt, whiskered muzzle. Eye and ear openings small. Forelimbs as blunt flippers, hind limbs lacking. Rounded back with no dorsal fin. Tail with concave trailing edge.

2. Orca (Killer Whale)   p. 122
Orcinus orca

Largest dolphin (blackfish) with diagnostic black and white markings. Up to 9.8m but with male much larger than female

(see 2a). Predominantly black with white underside expanding in a lobe onto the flanks. White oval behind eye and grayish saddle behind dorsal fin. Flippers large, broad, and rounded. Dorsal fin tall and curved back in female, in mature male (see 2b), very tall and straight.

3. Rough-toothed Dolphin   p. 122
Steno bredanensis

Small- to medium-sized, rather 'odd' looking dolphin. Generally confined to deep offshore waters. Up to around 2.8m. Dark above with paler flanks and white below. May have some pale blotching. Straight, sloping forehead continuous with slender beak. Large eyes. Dorsal fin prominent with concave trailing edge.

4. Striped Dolphin   p. 122
Stenella coeruleoalba

Small- to medium-sized, strikingly patterned dolphin. Up to 2.5m. Slim dolphin, perhaps most similar to Common Dolphin *Delphinus delphis*, but lacking hourglass pattern. Distinctively marked, dark above with pale gray flanks and tail stock. Slim beak is dark. Dark around eye with stripe extending to dark flippers and slender dark streak extending from eye to rear underside. Pale below. Dorsal fin dark and falcate, can be taller and more slender than shown.

_____ Plate 18_____

The Cape Hare, the Hyraxes, and the Porcupines

1. Cape Hare   p. 218
Lepus capensis
Very long-eared, large, rodent-like mammal. Ears with dark tips. Tail short and showing black-and-white in flight. Not an inhabitant of mountain slopes.

2. Rock Hyrax   p. 142
Procavia capensis
Stocky and robust mammal of mountainous terrain. Short-eared and tailless. Generally brownish, turning buffish below. Pale eyebrows indistinct. Sociable and active during the day.

3. Yellow-spotted Hyrax   p. 144
Heterohyrax brucei
Similar to previous species but generally smaller and slighter, grayish above with clearly demarcated white to cream underparts. Pale eyebrows distinct even at a distance.

4. Crested Porcupine   p. 212
Hystrix cristata
Unmistakable. Large rodent, far larger than any hedgehog, with prominent and elongated black-and-white spines. Southeastern Egypt. The Indian Crested Porcupine *Hystrix indica* very similar, but in Egypt, only in North Sinai.

①

Plate 19

The Antelopes
and the African Wild Ass

1. Scimitar-horned Oryx   p. 149
Oryx dammah

Large, very pale antelope with character-istic horns. Largely white with reddish chestnut, particularly on the neck. Both sexes with very long, slender, back-curved horns, but thicker and shorter in the male. Now globally extremely rare and almost certainly extinct in Egypt. Historically, from the Western Desert.

2. Addax   p. 147
Addax nasomaculatus

Smaller than previous species and with very different horns. Cream colored, desert antelope, though shade varies with age and time of year. Forequarters darker. Crown dark brown between horns. Younger animals darker, more reddish. Horns in both sexes rather slender and spiraled (straighter in young animals). Now globally extremely rare and almost certainly extinct in Egypt. Historically, from the Western Desert.

3. African Wild Ass   p. 138
Equus africanus

Very similar to domestic donkey. Egypt's only wild equid and only mistakable for feral donkeys. Note white muzzle, dark shoulder and dorsal stripe, and absence of stripes on legs, a feature common in feral donkeys. Only likely in extremely remote areas of the southern Eastern Desert.

Plate 20

The Gazelles

The gazelles *Gazella* spp. are distinctive, small, and slender ungulates unlikely to be confused with any other Egyptian mammals. The limbs are very slender, the body compact and often strongly marked. Males have well-developed horns; in females, horns are less developed and more slender. Key points to look for are the presence or absence of a flank stripe, and the strength of the facial markings. In adult males, horn structure can be important.

1. Dorcas Gazelle   p. 151
Gazella dorcas

Small and slender gazelle, by far the most widespread in Egypt. Coat fairly uniform, generally with no distinct flank stripe. Facial markings clear. Ears elongated and prominent. Horns in male annulated, moderately long, and turned in at tip, more slender in female.

2. Slender-horned Gazelle   p. 153
Gazella leptoceros

Slenderly built gazelle with no clear facial markings and no flank stripe. Pallid. With elongated and slender horns in male, not turned in, smaller in female. Only in the Western Desert.

3. Mountain Gazelle   p. 154
Gazella gazella

Rather large, slender gazelle. Pale red-brown with distinct darker flank stripe. Horns in male fairly robust and moderately lyrate, more slender in female. Ears prominent but less so than in Dorcas Gazelle. In Egypt, Sinai only.

① ② ③

_____ Plate 21 _____

The Barbary Sheep
and the Nubian Ibex

Large ungulates of mountainous desert regions. For details, see species descriptions.

1. Barbary Sheep   p. 158
Ammotragus lervia
Robust, short-limbed intermediate between sheep and goats. Southern deserts only. Both sexes golden tan in color with long fringes of lighter hair on forelegs and throat, longer and more elaborate in the male (illustrated) and lacking chin tuft. Horns ridged, but not knobbed, curving more out than back. Longer and thicker in the male. No contrasting leg markings.

2. Nubian Ibex   p. 156
Capra ibex
Robust, short-limbed wild goat. Mountains of the Eastern Desert and Sinai, not recorded west of Nile. Stocky, both sexes brown with fore- and hind limbs distinctively marked black and white. Chin tuft present in male. Black line along spine. Horns present in both sexes and are long and back-curved, much longer and heavier in the male, and heavily knobbed (illustrated).

Plate 22

The Jerboas, the Bushy-tailed Jird, and the Middle Eastern Dormouse

The jerboas have greatly elongated hind limbs and long tails with a distinct black tuft and white terminal tuft. The Bushy-tailed Jird *Sekeetamys calurus* has much less developed hind limbs, a proportionately shorter tail with a more extensive black tuft also white tipped. The Middle Eastern Dormouse *Eliomys melanurus*, too, has an extensive black tuft, normally tipped white, but differs in having a distinctive black facial mask.

1. Four-toed Jerboa p. 191
Allactaga tetradactyla
Distinctive small jerboa confined to coastal desert west of Alexandria. Typically long limbed, elegant, with long, black, tufted tail with white tip. Ears long and slender. In hand, can be distinguished by presence of a fourth, non-functional toe on hind foot.

2. Lesser Egyptian Jerboa p. 193
Jaculus jaculus
Small jerboa. Very widespread. Long, slender hind limbs with long, black-tufted tail with white tip. Ears shorter than in previous species. In hand, note three functional toes on hind foot. Hind foot shorter than 70mm.

3. Greater Egyptian Jerboa p. 195
Jaculus orientalis
Large jerboa, much larger than previous species and restricted to coastal desert west of Alexandria and Sinai. Ears are long but proportionately shorter than in the Four-toed Jerboa. Long, slender hind limbs with long, black-tufted tail with white tip. In hand, note three functional toes on hind foot. Hind foot longer than 70mm.

4. Bushy-tailed Jird p. 189
Sekeetamys calurus
Large, gerbil-like rodent of rocky slopes in the Eastern Desert and Sinai. Face without facial mask. Hind limbs shorter than in jerboa species. Tail largely blackish and bushy with white tip.

5. Middle Eastern Dormouse p. 210
Eliomys melanurus
Robust, rather squirrel-like rodent of northern Western Desert and Sinai. Tail shorter than head and body length, largely black, bushy with white tip. Black facial mask diagnostic.

⑤

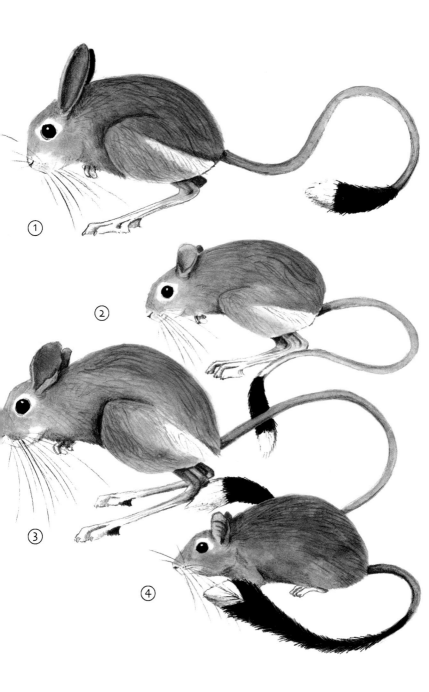

Plate 23

The *Gerbillus* Gerbils (1)

Very small to small desert and desert margin/oasis rodents. Brown to yellowish brown to orange above, but very variable between and within species, white below. Presence and extent of white supra/postorbital and postauricular patch and extent of white at the base of the tail important for identification (see text p. 163). All told from *Meriones* jirds by smaller size and by tail length being longer than head and body length (except for Simon's Gerbil *Dipodillus simoni*). Distinguished from jerboas by shorter hind limbs and lack of black and white tuft to the tail.

1. **Lesser Egyptian Gerbil**   p. 170
Gerbillus gerbillus
Small gerbil. Very widespread. Pale, yellow-orange to orange-brown above, white below. Much white on face and around eyes. Postauricular patch distinct. Rump patch large and white. Ear margin blackish. Tail bicolored with fairly distinct brush about one-third of tail length.

2. **Greater Egyptian Gerbil**   p. 175
Gerbillus pyramidum
Largest gerbil in Egypt. Very widespread. Rather variable. Orange-brown to brown above, often with darker dorsal stripe, white below. Desert specimens pale around eyes, those from the Delta/Valley less so. Rump patch absent in some populations. Tail variable, white below to base, or buff below at base. Brush moderately conspicuous, brown, about one-third of tail length.

3. **Giza Gerbil**   p. 164
Gerbillus amoenus
Small gerbil. Western Desert and southern Eastern Desert. Darkish yellow-brown above to yellow on flanks, white below. White supraorbital and postauricular patches distinct. Rump patch large and distinct. Ear tip pigmented. Tail bicolored, white below. Brush obscure, about one-third of tail length.

4. **Pygmy Gerbil**   p. 171
Gerbillus henleyi
Very small, Egypt's smallest gerbil. Widespread, though absent from much of Western Desert. Buff-brown above, paler on flanks, white below. White supraorbital and postauricular patches prominent. Rump patch white, prominent. Ear not pigmented. Tail bicolored, white below. Brush obscure, one-quarter or less of tail length.

Plate 24

The *Gerbillus* Gerbils (2)
The Western Desert Gerbils

1. Anderson's Gerbil p. 165
Gerbillus andersoni
Medium-sized gerbil. Northern Western Desert, Siwa, Delta south to the Fayoum. For *G. a. bonhotei* of North Sinai, see plate 25. Orange-brown above, with dorsal stripe to clear orange on flanks, white below. Whitish postorbital and postauricular patches small and indistinct. Rump patch small, white. Ears pigmented. Tail bicolored, white below except at base where buff, above as upper parts with scattered black hairs. Brush indistinct, brownish, and about one-quarter of tail length.

2. Pallid Gerbil p. 177
Gerbillus perpallidus
Medium-large, pale gerbil. Northeastern Western Desert including Wadi Natrun. Pale yellowish to reddish orange above. No dorsal stripe. White below. Postorbital and postauricular patches white and distinct. Rump patch white and distinct. Ears not pigmented. Tail bicolored, white below. Brush indistinct, yellowish brown, and about one-third of tail length.

3. Lesser Short-tailed Gerbil p. 178
Dipodillus simoni
Small gerbil with tail only slightly longer than head and body length. Mediterranean coastal desert west of the Delta. Yellowish brown above with clear yellow on flanks, white below. Postorbital patch whitish and indistinct. Postauricular patch whitish and large. Rump patch absent. Ears pigmented. Tail bicolored, buffish above, with scattered black hairs, white below. Brush virtually absent.

4. Large North African Gerbil p. 167
Gerbillus campestris
Medium-sized gerbil. Scattered records from throughout the Western Desert. Variable with four subspecies described from Egypt. Orange-brown to brown above with orange on flanks to foreleg, white below. Postorbital and postauricular patches pale, indistinct. Rump patch indistinct if rump hairs are banded white, or absent. Ears pigmented. Tail as upper parts above, variable below, white to brownish. Brush variably distinct to indistinct, one-third to one-half of tail length.

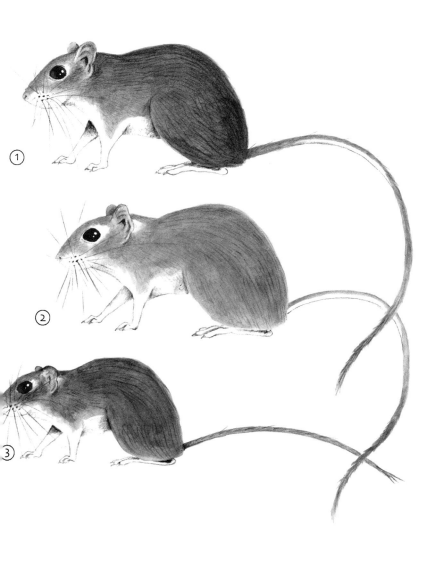

Plate 25

The *Gerbillus* Gerbils (3)·
The Eastern Desert and Sinai Gerbils

1. Mackilligin's Gerbil   p. 173
Gerbillus mackilligini
Small gerbil with long tail. Sudan Government Administration Area only. Dark yellow-brown above, sides paler to narrow yellow flank stripe, white below. Postorbital and postauricular patches whitish, indistinct. Rump patch absent. Ears pigmented. Tail variably bicolored, whitish to buff below, upper side darker than back and with blackish hairs scattered to tail base. Tail brush distinct, dark, and about one-half of tail length.

2. Wagner's Gerbil   p. 168
Gerbillus dasyurus
Medium-small gerbil. Northern Eastern Desert and Sinai. Yellow-brown above paling to narrow yellow flank stripe, white below. Postorbital and postauricular patches whitish and indistinct. Rump patch absent. Ears pigmented. Tail bicolored, whitish below. Tail brush yellowish brown, about one-half of tail length.

3. Anderson's Gerbil   p. 165
Gerbillus andersoni bonhotei
Medium-sized gerbil. Northeast Sinai only. As Anderson's Gerbil *G. andersoni* (see plate 24) but overall paler.

4. Baluchistan Gerbil   p. 174
Gerbillus nanus
Small gerbil. Sinai and possibly southern Eastern Desert. Yellow, sandy-brown above with buffish flanks. White below. Cheeks buff. Postorbital and postauricular patches white and distinct. White rump patch large and distinct. Only tip of ear pigmented. Tail bicolored, silvery-white below. Tail brush grayish, distinct, and about one-quarter of tail length.

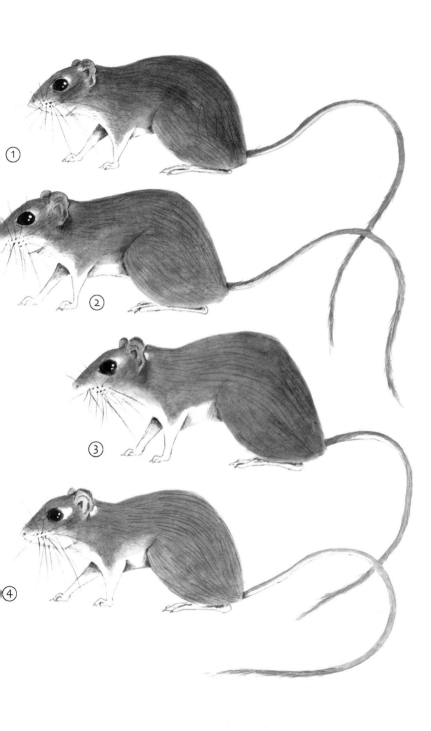

Plate 26

The Jirds

Desert rodents, yellow-brown to brown above, and white below. Superficially similar to the *Gerbillus* gerbils but may be distinguished from all the gerbils by larger size and, except from Simon's Gerbil *Dipodillus simoni*, by the tail being less than or equal to head and body length. The jirds are distinguished from the Fat Sand Rat *Psammomys obesus* by a relatively longer tail and white underparts. The tail is fully haired and the length and degree of tufting on the tail is valuable in distinguishing the species.

1. Tristram's Jird   p. 184
Meriones tristrami
Smallest Egyptian jird. Northeast Sinai only. Pale, dark yellowish-brown above with flanks yellowish orange. White below. Postorbital spot prominent. Postauricular spot prominent. Claws pale. Tail bicolored with underside of base orange tinged. Tuft blackish and inconspicuous, about one-quarter of tail length.

2. Negev Jird   p. 185
Meriones sacramenti
Largest Egyptian jird. Northeast Sinai only. Dark brownish jird with flanks clear cinnamon. White below. Supraorbital spot indistinct and small. Postauricular spot small, off-white. Claws pale. Tail indistinctly bicolored with black hairs scattered throughout length. Tuft blackish, distinct, and about one-third of tail length.

3. Silky Jird   p. 179
Meriones crassus
Large, pale jird. Widely distributed. Pale yellow-brown above with buff flanks. White below. Pale around eyes. Postauricular spot distinct and white. Claws pale. Tail distinctly or indistinctly bicolored. Tuft black, distinct, and about one-third to one-half of tail length.

4. Shaw's Jird   p. 182
Meriones shawi
Large, brownish jird. Coastal desert west of Alexandria. Yellowish brown above with yellow to orange on flanks. White below. Preorbital and postorbital spots distinct and grayish. Postauricular spot small and white. Claws pale. Tail uniform. Tuft blackish, distinct, and one-quarter of tail length.

5. Libyan Jird   p. 181
Meriones libycus
Large, brownish jird. Northern Western Desert to Wadi Natrun and the Fayoum. Yellowish brown above, with orange on flanks. White below. Grayish around eyes. Postauricular spot small and white. Claws dark. Tail fairly uniform with scattering of black hairs. Tuft blackish, distinct, and one-third of tail length. In field, tuft very prominent.

Plate 27

Unusual Rodents
The Fat Sand Rat, the Lesser Molerat, and the Fat-Tailed Jird

These three rodent species are not closely related but are sufficiently different from other Egyptian rodents (and, indeed, from each other) to not fit comfortably in other plates.

1. Fat Sand Rat   p. 187
Psammomys obesus
Robust, thick-tailed rodent typical of salt marsh habitat. One of the very few Egyptian rodents regularly encountered by day. Larger than Egyptian jirds and gerbils and heavier in build. Ears proportionately small and relatively short tail is robust with dark tip. Not bicolored.

2. Lesser Molerat   p. 214
Spalax leucodon
Robust rodent without external eyes, ears, or tail. Unmistakable. Lack of external features makes confusion with any other Egyptian mammal highly unlikely. The Lesser Molerat is largely fossorial and the pyramidal mounds marking the molerat's burrows are far more often encountered than the animal itself. At close quarters, note whitish head markings. North coast only from Alexandria west.

3. Fat-tailed Jird   p. 186
Pachyuromys duprasi
Superficially similar to other Egyptian jirds but peculiar club-shaped, short tail unique and diagnostic. Overall pale, brownish above, white below. Unmistakable.

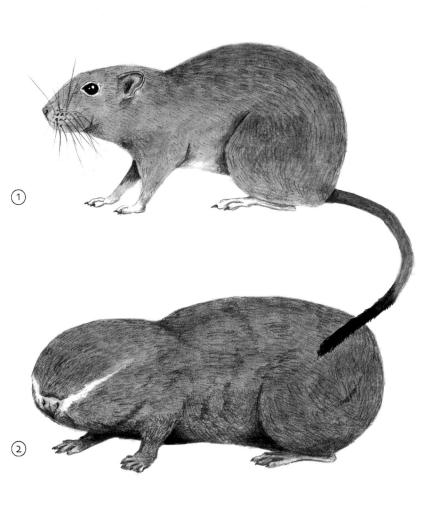

Plate 28

The Mice

1. Cairo Spiny Mouse   p. 198
Acomys cahirinus
Highly variable mouse with spines on dorsum extending from behind shoulder to base of tail. Urban populations tend to be darker (see 1b) than rural and desert ones (see 1a) but much variation. Ears large. Tail roughly length of head and body. Sparsely haired and clearly scaled.

2. Golden Spiny Mouse   p. 200
Acomys russatus
Golden color and extensive spines distinctive. Spines more extensive than in previous species, extending onto the head. Small white spot below eyes. Ears propor-

tionately smaller than in previous species. Tail blackish and shorter than head and body length. Restricted to Eastern Desert and South Sinai.

3. House Mouse   p. 202
Mus musculus
Small, uniform mouse lacking any distinctive markings. Only Egyptian mouse with no spines. Color variable but underside always paler. Tail roughly equal to head and body length, scales distinct. Distinguished from all gerbils by shorter, scaled tail with no trace of tuft, smaller eyes, and more pointed muzzle. Commensal.

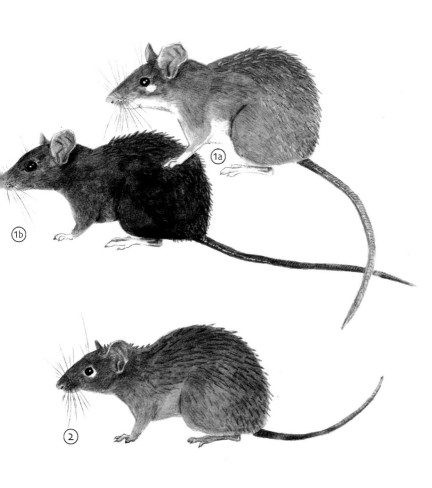

_____ Plate 29 _____

The True Rats, the Nile Kusu, and the Bandicoot Rat

1. House Rat p. 205
Rattus rattus
Large, long-tailed rat. Variable. Upper parts beige through brown to almost black, whitish to grayish below. Coat coarse. Eyes prominent. Snout pointed. Ears large and rounded. Tail longer than head and body length, slender, virtually naked with generally more than 200 rings. No tuft.

2. Brown Rat p. 206
Rattus norvegicus
Large, robust rat, more heavily built than the previous species. Upper parts dark brown, graying on the flanks, whitish to grayish below. Coat coarse. Eyes less prominent and ears smaller than in the previous species. Snout rather blunt. Tail normally shorter than head and body length, thick and heavy, with generally less than 200 rings. No tuft.

3. Nile Kusu p. 203
Arvicanthis niloticus
Large, rather slim, large-headed rat. Grizzled black-and-yellow above, often with dark, dorsal stripe. Tinged orange toward rump. Underparts whitish to grayish. Coat coarse. Ears small and rounded. Tail slim, about 80% of head and body length, fully furred (rings not visible). Bicolored. No tuft.

4. Bandicoot Rat p. 207
Nesokia indica
Large, robust, short-tailed rat. Upper parts brown, fading to buffish along flanks and underside. Big-headed, often with white throat patch. Ears rather large and virtually naked. Tail thick, sparsely haired, and shorter than head and body length. No tuft. Claws long.

④

_____ Plate 30 _____

Selected Tracks and Trails (1)
The Carnivores

1. Red Fox   p. 74
Vulpes vulpes
Track and trail shown. Track similar to that of a dog but smaller and slimmer. Note small size of interdigital pad that is no larger than the other pads. This feature is distinctive of all foxes. Claw marks may not show. In soft sand, the pad pattern may not be as distinct and may be further obscured by interdigital fur. Other foxes smaller.

2. Egyptian Mongoose   p. 89
Herpestes ichneumon
Track with four toes though claws will only show in soft ground. Interdigital pad relatively large.

3. Jackal   p. 70
Canis aureus
Larger than fox track and with the interdigital pad significantly larger than the other pads. Claws not always clear, especially on soft sand. Probably not safely distinguishable from the tracks of a similarly-sized domestic dog.

4. Egyptian Weasel   p. 81
Mustela subpalmata
Very small—smallest Egyptian carnivore. In good conditions, note five toes. Claws tiny and may not make an impression.

5. Marbled Polecat   p. 85
Vormela peregusna
Only likely in the very northeast of Sinai. Larger than previous species, but likewise fifth digit may not be apparent and claws obscure.

④

Egyptian Weasel
Mustela subpalmata

①

Red Fox
Vulpes vulpes

50mm

50mm

Egyptian Mongoose
Herpestes ichneumon

Jackal
Canis aureus

50mm

Red Fox trail

Marbled Polecat
Vormela peregusna

Plate 31

Selected Tracks and Trails (2)
The Ungulates

1. Nubian Ibex
p. 156
Capra ibex
Track and trail shown. Track very similar to that of domestic goat but larger, especially in adult male. Dewclaws do not leave an impression.

2. Dorcas Gazelle
p. 151
Gazella dorcas
Track noticeably narrow with hoof tips relatively pointed. More slender and pointed than the tracks of either domestic sheep or goat. Dewclaws do not leave an impression.

3. Barbary Sheep
p. 158
Ammotragus lervia
Track similar to track of Nubian Ibex but slightly larger and broader, more splayed. Male larger than female. Dewclaws do not leave an impression.

① Nubian Ibex
Capra ibex

75mm

② Dorcas Gazelle
Gazella dorcas

50mm

Nubian Ibex
trotting

③ Barbary Sheep
Ammotragus lervia

75mm

Plate 32

Selected Tracks and Trails (3)
The Rodents and the Cape Hare

1. Cape Hare   p. 218
Lepus capensis
Track and trail shown. Track very distinctive with very elongated hind foot and almost circular and much smaller forefoot. In loose sand, foot pads unlikely to be clear. Trail shown when hopping. When moving at speed, i.e., bounding, hind feet appear before forefeet and stride is far longer.

2. Crested Porcupine   p. 212
Hystrix cristata
Very distinctive track. Hind foot longer than forefoot. Large, largest Egyptian rodent, with broad pads. Hind foot with five toes, though fifth may not be clear,

• forefoot with four toes of more or less equal size. In Sinai, possible Indian Crested Porcupine *Hystrix indica* almost identical.

3. House Rat   p. 205
Rattus rattus
Small tracks but clearly larger than those of mouse species. Brown Rat *Rattus norvegicus* very similar but slightly larger and with hind foot more elongated. Five toes on hind foot, four on forefoot.

4. Cairo Spiny Mouse   p. 198
Acomys cahirinus
Very small tracks, smaller but otherwise similar to tracks of *Rattus* spp. Five toes on hind feet, four on forefeet.

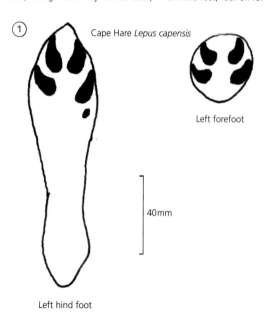

Cape Hare *Lepus capensis*

Left forefoot

40mm

Left hind foot

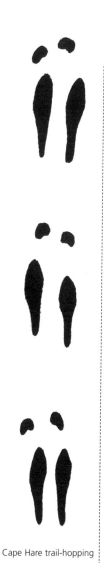

Cape Hare trail-hopping

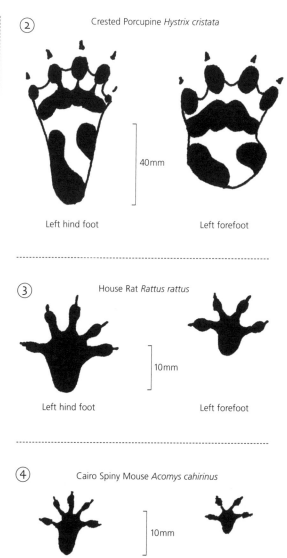

② Crested Porcupine *Hystrix cristata*

40mm

Left hind foot   Left forefoot

③ House Rat *Rattus rattus*

10mm

Left hind foot   Left forefoot

④ Cairo Spiny Mouse *Acomys cahirinus*

10mm

Left hind foot   Left forefoot

Glossary

Adult: Sexually mature animal capable of breeding though may or may not be doing so due to external influences, e.g., social hierarchies.

Annulated: Ringed, as in the horns of gazelles and antelopes.

Auricular/postauricular: Of the ear/behind the ear.

Baleen: The filtering plates in the upper jaws of the RORQUALS and certain other whales, such as the right whales.

Beak: The snout of many dolphin and whale species.

Biome: A naturally occurring group of plants and animals adapted to the particular environmental conditions in which they occur.

Blow: See SPOUT.

Bow riding: The habit of certain dolphin species to ride in the pressure wave of a ship or boat.

Breaching: The habit of whales and dolphins of leaping completely out of the water and slamming back in.

Brush: The tail of a fox.

Calcar: A spur, which supports the INTERFEMORAL MEMBRANE, arising from the ankle in many bats.

Canine: Next tooth after the incisors (unless absent). In many mammals, long and pointed.

Caprines: Goat-like animals.

Carnassial: Specifically in the Carnivora. The first lower molar and last upper premolar that together form a cutting or shearing edge.

Cetacean: Mammal entirely restricted to life in fresh or sea water but **not** a SIRENIAN, i.e., whale, dolphin, or porpoise. A member of the order Cetacea.

CITES: Convention on the International Trade in Endangered Species.

Class: In taxonomy, the grouping between phylum and order. E.g., class Mammalia.

Clinal: Not relating to a distinct subspecies but rather a continual change over a certain geographical range.

Cloaca: Single opening of the reproductive, urinary, and alimentary systems.

Commensal: Associated with man though capable of independent existence, e.g., Brown Rat and House Mouse.

Crepuscular: Active at twilight or dusk.

Dewclaws: Non-functional fifth claw of certain carnivores, e.g., dogs. Also used for the non-functional digits of other mammals.

Diurnal: Active by day.

Dorsal: Relating to the back or upper parts.

Dorsal fin: The fin on the back of many CETACEANs.

Dorsum: The upper parts or back.

Drey: Specifically a squirrel's nest, but also applied to the nests of certain other small rodents.

Echolocation: Particularly with bats and CETACEANs, the 'sense' that locates objects or prey by reflected sound waves.

Endemic: Specific to a particular region, area, or country.

Extant: Still existing. C.f. EXTINCT.

Extinct: No longer existing, e.g., Bubal Hartebeeste.

Extralimital: Beyond the range covered by any particular book or work.

Falcate: Referring to the DORSAL FIN of a CETACEAN being sickle shaped and curved back.

Family: In taxonomy, the grouping between ORDER and GENUS.

Feral: Of domestic animals, e.g., cats and dogs, living wild.

Flank: The side.

Flipper: The forefin ('arm') of CETACEANs.

Flukes: The tail fins of CETACEANs.

Forearm: Especially in bats. The measurement between the wrist and elbow.

Fossil: Organic remains preserved as mineral or rock.

Fossorial: Burrowing, e.g., Molerats.

Genus: In taxonomy, the grouping between family and species.

Gestation: Period between conception and birth.

Gular sac: Throat sac containing glandular secretions in certain bats.

Hamada: Stony desert with rock fragments covering sedimentary rock base.

Herbivore: An animal with a diet consisting mainly of plants.

Hibernation: Winter inactivity. C.f. AESTIVATION.

Home Range: Area within which an animal is found. This area may or may not be defended. C.f. TERRITORY.

Hybrid: The (generally infertile) offspring of two different species, e.g., Mule or Hinny.

Incisor: Any of the front teeth coming between the canines (if present).

Insectivore: An animal that feeds largely on non-vertebrates. Taxonomically a member of the order Insectivora.

Interdigital pad: The 'palm' of a paw print.

Interfemoral membrane: In bats, the flight membrane between each of the hind legs incorporating the tail

Introduced: Not native. Has been brought to an area by intentional or unintentional human agency.

Invertebrate: Broadly any animal that does not possess a backbone.

Iris: The part of the eye encircling the pupil that may be distinctively colored.

IUCN: World Conservation Union.

Juvenile: Beyond infant yet not fully adult.

Keel: In CETACEANs, the root of the tail.

Kit: The young of certain carnivores, e.g., genets, weasels, etc.

Kingdom: In taxonomy, the grouping of like PHYLUMs, e.g., the Animal Kingdom.

Lactation: Characteristic of mammals—the secretion of milk from mammary glands.

Lancet: Erect part of the noseleaf of, e.g., horseshoe bats below the SELLA.

Lappets: Flaps of bare skin, generally around the facial region.

Latrine: Habitual place of defecation.

Lobtailing: The slapping of FLUKES against the water surface by CETACEANs.

Logging: In CETACEANS lying at the surface lazily, a school often orientated in the same direction.

Lyrate: Lyre-shaped. In reference to the horns of certain ungulates, e.g., Mountain Gazelle *Gazella gazella*.

Mammary glands: The glands that secrete milk. C.f. LACTATION.

Mammal: A member of the ORDER Mammalia possessing MAMMARY GLANDS.

Melanistic: A black color phase. C.f. the urban Cairo Spiny Mouse.

Melon: The bulbous forehead of many smaller CETACEANs.

Migration: Seasonal (generally) movement of a population in response to feeding or breeding stimuli.

Molar: Chewing teeth after the premolars.

Nocturnal: Active by night.

Nominate: The subspecies of a species that shares the species name as the trinominal, e.g., *Gerbillus gerbillus gerbillus*.

Noseleaf: The fleshy embellishments on the nose of certain bat species, associated with ECHOLOCATION.

Olfactory: Relating to the sense of smell.

Omnivore: An animal that feeds on a wide variety of food, both plants and animals.

Order: In taxonomy, the grouping between CLASS and FAMILY.

Orographic: Of rainfall, that which is caused by mountain ranges or other dominant physical features.

Palearctic: See WESTERN PALEARCTIC.

Papillae: Nipples or nipple-like protuberances.

Parasitism: Relationship whereby one species obtains benefit by living at the cost of another (generally larger) species.

Pelagic: Of the open sea.

Phylum: In taxonomy, the grouping between KINGDOM and CLASS.

Pod: A small group of whales or dolphins, possibly of related individuals. See SCHOOL.

Polygamous: In breeding, a system where one individual (in the strictest sense, male) has more than one mate.

Post-calcarial lobe: In bats, the lobe of skin present in some species on the outside edge of the CALCAR.

Predator: An animal that preys on other live animals.

Refection: Process in which food is eaten a second time on defecation, characteristic particularly of certain shrew species.

Relict: In the biological sense a species that is of particularly anti-quated origin.

Riverine: Of freshwater habitat along rivers.

Rorqual: A whale, often very large, of the genus *Balaenoptera,* pop-ularly including the Humpback Whale (*Megaptera novaeangliae*).

Ruminant: A mammal (often large herbivore) adapted to eating food twice, the second time after regurgitation and 'chewing the cud'.

Rut: The seasonal, aggressive, sexual behavior of male ruminants.

Sacrophytic mange: A disease common amongst certain mammals, e.g., foxes, leading to a severe lack of condition and extreme dis-comfort.

School: Popular collective noun for dolphins and whales. Larger than pod.

Sella: Nasal process of the horseshoe bats that sticks out above and overhangs the nasal horseshoe.

Sibling: An individual that shares a common parent or parents with those of the same litter or litters.

Sirenian: In Egypt a Dugong *Dugong dugon.* Elsewhere a Dugong or manatee species.

Sonar: See ECHOLOCATION.

Splashguard: Elevated area immediately before the blowhole of baleen whale species.

Species: The fundamental taxonomic division. The grouping between genus and subspecies. In theory the highest level at which two individuals can produce fertile offspring. In practice this is becoming increasingly clouded.

Spoor: The tracks, trails, or droppings of an animal.

Spout: The water vapor and droplets ejected as a large whale exhales on the surface.

Spyhopping: When a CETACEAN raises its head vertically out of the water to below the level of the eye to scan the surroundings.

Stranding: The beaching of a whale or dolphin.

Sub-adult: General term pertaining to individuals that are not juve-nile but not yet fully adult.

Subspecies: A particular subset or group of a species that while fully capable of successful reproduction with the main SPECIES is significantly isolated geographically or ecologically.

Taxonomy/taxonomics: The science that classifies all living organisms.

Terminal tuft: The tuft of elongated hairs at the tip of many mammals' tails that is often distinctively patterned in black and white

Territory: An area that is defended by an animal or by a group of animals. May or may not be the HOME RANGE. The territory of any individual may overlap that of any other individual, e.g., a male's territory may overlap the territory of several females but he will not tolerate another male within its boundaries.

Tragus: Especially in bats. A flap of skin in the ear opening, often of distinctive shape.

Ultrasound: Sound above the normal human range, used by, e.g., bats.

Ungulate: Hoofed animals belonging to several ORDERs. In Egypt, the Artiodactyla, the Perrisodactyla, and the Hyracoidea.

Ventral: Referring to the underparts.

Ventral fins: Fins on the anterior part of the body present in many fish but never in sea mammals.

Vertebrate: In taxonomic terms, any member of the subphylum Vertebrata of the PHYLUM Chordata. Broadly, an animal with a backbone.

Vibrissae: Commonly known as whiskers. The long, sensory hairs around the snout of many mammals.

Vulpine: Of foxes or fox-like.

Western Palearctic: Faunal region encompassing Europe east to the Urals, North Africa, and much of the Middle East.

Selected Bibliography

Anderson, J. 1902. *Zoology of Egypt: Mammalia* (completed by W. E. De Winton). London: Hugh Rees Ltd.

Bagnold, R. A. 1987. *Libyan Sands: Travel in a Dead World.* London: Michael Haag Ltd.

Baha El Din, S. M. 1999. *Directory of Important Bird Areas in Egypt.* Cairo: Birdlife International. Printed and distributed by Palm Press.

Brown, R. W., Lawrence, M. J. and Pope, J. 1992. *Animals: Tracks, Trails and Signs.* London: Hamlyn.

Burton, J. A. and Pearson, B. 1987. *Collins Guide to the Rare Mammals of the World.* London: Collins.

Carwardine, M. 1995. *Whales, Dolphins and Porpoises.* London: Dorling Kindersley.

Dixon, A. and Jones D., eds. 1988 *Conservation and Biology of Desert Antelopes.* London: Christopher Helm.

Edwards, A. J. and Head, S. M. 1987. *Key Environments: Red Sea.* Oxford: Pergamon Press with the IUCN.

Fenton, M. B. 1998. *The Bat: Wings in the Night Sky.* Shrewsbury: Swan Hill Press.

Ferguson, W. W. 1981. "The Systematic Position of *Gazella dorcas* (Artiodactyla: Bovidae) in Israel and Sinai." *Mammalia* 45 (4): 459–465.

Flower, S. S. 1932. "Notes on the Recent Mammals of Egypt, with a List of the Species Recorded for that Kingdom." *Proceedings of the Zoological Society of London*: 368–450.

Goodman, S. M. 1985. "The Occurrence of *Crocidura floweri* in Wadi Natrun, Egypt." *Mammalia* 53 (1): 134–135.

Goodman, S. M. and Meininger P. L. 1989. *The Birds of Egypt.* Oxford: Oxford University Press.

Hafez, S. 1993 *A Checklist of the Mammals of Egypt.* Cairo: Unpublished Report for the IUCN Monitoring Center.

Haltenorth, T. and Diller, H. 1980. *A Field Guide to the Mammals of Africa including Madagascar.* London: Collins.

Harrison, D. L. and Bates P. J. J. 1991. *The Mammals of Arabia.* Sevenoaks: Harrison Zoological Museum.

Hoath, R. C. "Mammal Mania [Botta's Serotine Bat]." *Egypt Today* Jan. 2006: 72–72.

———. "Some Like It Hot [Humpback Whale]." *Egypt Today* May 2008: 60–61.

Hobbs, J. 1990. *Bedouin Life in the Egyptian Wilderness.* Cairo: American University in Cairo Press.

———. 1994. *St. Katherine's Monastery.* Cairo: American University in Cairo Press.

Horwood, J. 1987. *The Sei Whale: Population Biology, Ecology and Management.* London: Croom Helm.

Houlihan, P. F. 1996. *The Animals of Ancient Egypt.* Cairo: American University in Cairo Press.

Khalil, R and Aly, D. 2000. *Egypt's Natural Heritage.* Cairo: Tourist Development Authority.

Kingdon, J. 1991. *Arabian Mammals: A Natural History.* London: Academic Press.

———. 1990. *Island Africa.* London: Collins.

———. 1997 *The Kingdon Field Guide to the Mammals of Africa.* San Diego: Academic Press Natural World.

Larsen, T. B. 1990. *The Butterflies of Egypt.* Cairo: Apollo Books/American University in Cairo Press.

Macdonald, D. 1985. *The Encyclopaedia of Mammals.* Oxford: Guild Publishing.

Macdonald, D. and Barrett, P. 1993. *Collins Field Guide to the Mammals of Britain and Europe.* London: Harper Collins.

McKinnon, M. 1990. *Arabia: Sand, Sea, Sky.* London: Immel Publishing.

Mendelssohn, H. et al. 1987. "On the Occurrence of Blanford's Fox in Israel and South Sinai." *Mammalia* 51 (3): 459–462.

Meininger, P. L. and Atta, G. 1994. *Ornithological Studies in Egyptian Wetlands 1989/90*. Netherlands: FORE–Report No. 94-01.

Nowak, R. M. ed. 1999. *Walker's Mammals of the World*. Baltimore: Johns Hopkins University Press.

Osborn, D. J. and Helmy, I. 1980. "The Contemporary Land mammals of Egypt (Including Sinai)." *Fieldiana: Zoology*. New Series, No 5.

Qumsiyah, M. B. 1985. "The Bats of Egypt." *Special Publication Museum Texas Tech. University* 23: 1–102.

Reeves, R. R. et al. 2002. *Sea Mammals of the World*. London: A&C Black.

Saleh, M. A. and Basuony. 1998. "A Contribution to the Mammalogy of the Sinai Peninsula." *Mammalia* 62 (4): 17–35.

Stuart, C. and Stuart, T. 1988. *Field Guide to the Mammals of Southern Africa*. London: New Holland.

Wassif, K. 1995. *Guide to the Mammals of Natural Protectorates in Egypt*. Cairo: Egyptian Environmental Affairs Agency (National Biodiversity Unit).

Wassif, K. and H. Hoogstraal. 1953. "The Mammals of South Sinai, Egypt." *Proceedings of the Egyptian Academy of Science* 9: 63–79.

Watson, L. 1981 *Sea Guide to the Whales of the World*. London: Hutchinson.

Zalat, S. and Gilbert, F. 1998 *A Walk in Sinai: St Katherine to Al Galt Azraq*. Cairo: Published by the authors.

Index of Scientific Names

Figures in **bold** type refer to the plate number of the main species illustration.
Where a second plate number is given, this indicates the track and trail illustrations.

Index of English Names

Figures in **bold** type refer to the plate number of the main species illustration.
Where a second plate number is given, this indicates the track and trail illustration.

Index of Arabic Names

Figures in **bold** type refer to the plate number of the main species illustration.
Where a second plate number is given, this indicates the track and trail illustration.